D1174548

# Wireless
# Security
# Handbook

# Wireless Security Handbook

## Aaron E. Earle

Auerbach Publications
Taylor & Francis Group
Boca Raton   New York

Published in 2006 by
Auerbach Publications
Taylor & Francis Group
6000 Broken Sound Parkway NW, Suite 300
Boca Raton, FL 33487-2742

International Standard Book Number-10: 0-8493-3378-4 (Hardcover)
International Standard Book Number-13: 978-0-8493-3378-1 (Hardcover)
Library of Congress Card Number 2005049924

### Library of Congress Cataloging-in-Publication Data

Earle, Aaron E.
    Wireless security handbook / Aaron E. Earle.
        p. cm.
    Includes bibliographical references and index.
    ISBN 0-8493-3378-4 (alk. paper)
    1. Wireless LANs--Security measures. 2. Wireless communication systems--Security measures. I. Title.

TK5105.78.E23 2005
005.8--dc22

2005049924

Taylor & Francis Group
is the Academic Division of Informa plc.

**Visit the Taylor & Francis Web site at
http://www.taylorandfrancis.com** .

**and the Auerbach Publications Web site at
http://www.auerbach-publications.com**

# Contents

# Preface

This book was written to give the reader a well-rounded understanding of wireless network security. It looks at wireless from multiple perspectives, ranging from auditor, to security architect, to hacker. This wide scope benefits anyone who has to administer, secure, hack, or participate on a wireless network. Going through this book, the reader will see that it tackles the risk of wireless from many angles. It goes from a policy level to mitigate certain risks that wireless brings. It talks about the most cost-effective solutions to deploy wireless across a large enterprise. It talks about financial and technical controls that one can apply to reduce any unforeseen risk involved in a large wireless project. It covers the technical details of how to design, build, and hack almost all wireless security methods.

The wide scope of knowledge that this book brings will help one become acquainted with the many aspects of wireless communications. This book also has career advancement in mind by covering all the objectives of the three widely upheld wireless certifications currently on the market. These certifications are administered by Planet3 Wireless and Cisco Systems. The focus of this book is on wireless local area networking technologies to meet these objectives, although this book looks at the security of almost all mobile communications. So if you are interested in obtaining a certification or just a deep knowledge of wireless security this book is for you.

# Acknowledgments

I would like to thank many people who over the years have helped me get to where I am today. Great wisdom comes from one who knows that it is not what you do to advance, but rather what the people below you do to push you in that direction. I would like to thank my family and friends who have supported me throughout this endeavor, and my girl-friend Clare who did not complain about the long hours away from her spent writing this book. I would like to thank my father Douglas R. Earle, who purchased my first computer for me; my friend Justin Peltier, who gave me the "I can do it, you can do it" mentality; and my friend Paul Immo, who saw my passion for technology and helped me achieve my goals around education and certification. I would also like to thank my friend Jeremy Davison for allowing me to forget altogether about computers, networking, security, and technology and just have fun every now and then.

# Chapter 1

# Wireless Network Overview

This chapter looks at radio frequencies (RF) in general. The goal of this chapter is to gain a general understanding of RF. This allows us to see what issues are inherent in all wireless communications, whether it is a cell phone or an 802.11g laptop. This knowledge can help us troubleshoot RF networks and understand what can and cannot be fixed. After reading this, we look at the many types of interference that affects all wireless communications. Once an understanding of interference is achieved, we look at modulation. We discuss the different types of modulation used on wireless networks and how each of them works. The final section of this chapter addresses the many wireless groups that create and regulate the way we use wireless communications.

## 1.1 RF Overview

What are radio frequencies, and where did they come from? Radio frequencies are nothing more than power, in the form of an alternating current created by an electrical device that passes though wiring and out an antenna. The antenna then radiates this power, creating radio waves that travel across the air in all directions until the waves become so minute that one cannot detect them. Heinrich Hertz discovered radio transmission in the late 1880s; he expounded on James Clerk Maxwell's research on the electromagnetic theory of light. Hertz found that by using a strong electrical signal it was possible to send that signal through nonconductive material; later, the notion of such material went out the door when Hertz discovered that the signal could conduct through the air. This is how radio signals and thus wireless communications were born.

As the radio waves travel across the air, a receiving antenna can take the wave and convert it back to an electrical signal. This signal would be the same as the one originally created by the sending electrical device. The way wireless propagates itself is very similar to dropping a stone into a large body of water. Once the stone hits the water, ripples are created, moving in all directions until the ripples are so minuscule that they no longer can be seen or detected.

Electromagnetic waves are produced by the motion of electrically charged particles. These waves are also called *electromagnetic radiation* because they radiate from the electrically charged particles. All wireless devices have some form of electromagnetic waves. All these waves are part of the electromagnetic spectrum; this spectrum has all types of electromagnetic radiation classified. Although the size of this spectrum is infinite, the size of the radio portion is limited to around 100 kHz to 300 GHz. The waves discussed herein are mostly based in the microwave section of the radio spectrum. The larger an electromagnetic wave, the further it will travel. The fact is that when you look at radio waves, the amount of information being sent is small, and therefore the frequency used is also small. A small frequency signal has a very large wave. A radio wave, like the ones one picks up on a car radio, can be thought of as about the size of an adult elephant.

Now look at an x-ray wave. This is very high on the radio spectrum, so it will have a large amount of data traveling down a small wavelength. This wave might be as small or smaller than a single atom. This smaller x-ray wave will not travel as far as the radio wave because of its limited size.

In discussing frequency, one must understand how to measure it. When looking at a wave traveling in time, one can see the amount of times a signal wave is completed from an upper crest to its lower crest. Each time this is completed, it is a single cycle. When one measures the total amount of wave cycles in a particular amount of time, one gets a frequency. In general, one takes the amount of cycles in a single second, giving the hertz (Hz). In the case of wireless networks, this amount is so large that it is measured in gigahertz (GHz), which is one billion hertz.

When talking about power and wireless, there are a number of values commonly used to measure wireless power. The first value to look at is the Watt, the rate at which a device converts electrical energy into another form of energy, such as light, heat, or — in this case — a wireless signal. The Watt can be measured in a number of ways, depending on how high or low a value it is compared to a single watt. What this means is if one has a value much greater than a single watt, maybe somewhere around 1000 watts, one would have a kilowatt (kW). This is because a kilowatt represents 1000 watts. Now, if one has less than a single watt, then one has a milliwatt (mW), which is 1/1000 of a watt. The milliwatt is the primary watt designation in relation to wireless local area networking.

The next term is the decibel. A decibel (or dB) is a mathematical — or, to be specific, a logarithmic ratio — that indicates the relative strength of a device's electric or acoustic signal to that of another. This can be used by itself, although it is mostly used with a unit of measurement. Looking at wireless, the most common units of measurement used with the decibel are the milliwatt (dBm), the forward gain of an antenna compared to an imaginary isotropic antenna (dBi), and the forward gain of an antenna compared to a half-wave dipole antenna (dBd). Wireless networks are measured in decibel strength compared to one milliwatt. In wireless local area networking (WLAN), dBi and dBd are commonly used and a formula is often needed to convert these two expressions into each other so they an be correctly compared. Chapter 8 goes into greater detail about both isotropic and dipole antennas and power measurement. Until then, just remember that these two figures are the most commonly used measurements of wireless power.

When discussing bandwidth, most computer people associate it with network performance. In the wireless world, bandwidth has a slightly different meaning. The meaning we are looking for in relation to wireless has to do with the size or the upper and lower limit to the frequency we are using. When we compare frequency and bandwidth, we see that frequency is a specific location on the electromagnetic spectrum compared to wireless bandwidth, which is the range between two frequencies. A single channel on the 2.4-GHz frequency has a channel bandwidth of 20 MHz. This is an example of wireless bandwidth. Looking at network performance bandwidth, one would identify it as the following: the network WAN connection only has a bandwidth of 1.5 megabytes (MB).

## 1.2 Wireless Signal Propagation

When radio waves travel in the air, many things affect their quality, thus prohibiting them from actually transmitting their intended signals. Interference is one of the oldest and most difficult problems facing every type of wireless communication. This interference has caused such a design

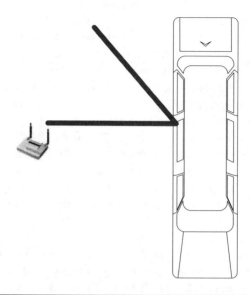

**Figure 1.1    Reflection.**

challenge throughout history that many governments from around the world have had to step in to make certain frequencies restricted from use. Restricting this use prevents interference caused by other wireless devices and makes for cleaner airwaves.

What happens to radio waves when interference affects their direction, influencing their signal clarity? Depending on what caused the interference, different common effects can occur. When the interference consists of certain objects, there are a number of well-documented, specifically proven results. When radio waves hit an object, they will bounce just like a child's ball. They also have the ability to pass through some objects just as a ghost would. Being able to understand when each of these occurrences takes place is critical to understanding the operation of wireless.

### 1.2.1 Reflection

Reflection takes place when an electromagnetic wave impacts a large, smooth surface and bounces off. This can happen with large surfaces such as the ground, walls, buildings, and flooring. After reflection takes place, radio waves often radiate in a different direction than originally intended. As one can see in Figure 1.1, the signal has a main pathway that intersects with the object. As it hits the object, it bounces off and heads in a different direction. This reflecting action lowers the signal

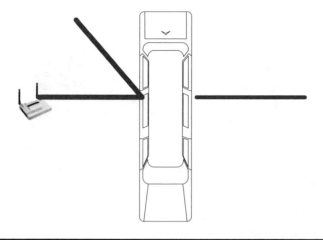

**Figure 1.2    Refraction.**

strength as it bounces off objects. Predominantly, the signal will pass through an object rather than bounce off of it. Reflection is one of the least obstructing interference types. This is because, for the most part, the signal remains whole; however, it moves in a different direction after it is reflected. Moreover, some of the other types of interference types will severely impact the signal's quality.

## 1.2.2 Refraction

When a signal reflects off an object and passes through it at the same time, one obtains what is called refraction (see Figure 1.2). RF is very stubborn; it goes places one does not want it to. Walls, buildings, or floors that should reflect the signal often do not; RF waves have a tendency to penetrate these objects instead. Once the signal has penetrated through these obstacles, it now has a degraded signal strength, which prevents it from reaching as far as it could have before the refraction. This is why reflection is not as bad an inherent interference as refraction. When a signal is reflected, most of the signal quality and strength is reflected with it. Refraction takes place when the signal has a portion of it penetrating and a portion of it reflecting. When this happens, the quality and strength are greatly deteriorated.

## 1.2.3 Diffraction

Diffraction, which is similar to refraction, describes what a signal does when it encounters an object. In diffraction, after the signal makes it

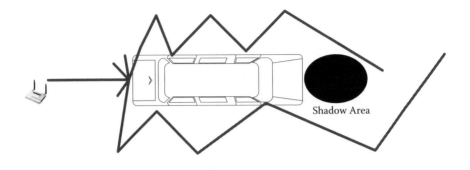

**Figure 1.3   Diffraction.**

around the object, we often get a shadow area. This is because the signal will bend around objects as best it can; but without being able to penetrate through the object, there is a dead spot created directly behind the object. Diffraction, unlike refraction, describes how the signal beams around objects instead of passing through them. People tend to get the two confused. In diffraction, shadow areas are created when an object will not allow refraction to occur. To picture this, see Figure 1.3, which shows the signal bending around the object; in doing so, it creates a shadow area directly behind the object. If refraction took place instead of diffraction, then the shadow area would not exist. This is because with refraction, the signal would bleed through the object and be present directly behind it. Some of the confusion around diffraction and refraction has to do with receiving a signal directly behind an object that the signal cannot penetrate. There are cases where this is true. It is possible for a signal to be unable to refract through an object but still be able to reflect enough times between different objects to make it around the main object.

## *1.2.4 Scattering*

Scattering (Figure 1.4) occurs when the RF signals encounter a rough surface or an area with tightly placed objects. The best way to understand scattering is to think of an automobile assembly line. In this scenario, one would see large amounts of robotic arms, raised metal-screened catwalks, pallets of metal doors, and many other objects. All these objects make the signal split into smaller signals, reducing the original signal's strength. The main signal enters this area and reflects off the small metal objects and ping-pongs, thus creating more and more signals. Over time, this makes the main signal so scattered that its original strength diminishes. This is because when scattering takes place, the signal is equally divided among the many waves bouncing around the tightly packed area. On top of the signal strength reduction, this type of interference can cause

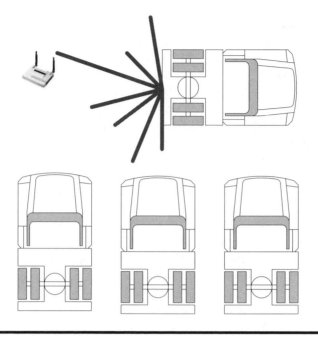

**Figure 1.4   Scattering.**

problems in receiving a signal. This is due to the fact that when multiple signals arrive at the receiver at the same time, it is difficult to correctly understand either of them.

### 1.2.5 Absorption

Just by the name, one can probably figure this one out. When a signal hits certain objects — mostly water-based objects such as trees, cardboard, or paper objects — the RF signal actually is absorbed into the object. This one interference problem plagues point-to-point or point-to-multipoint bridge operations. Trees having a large amount of water in them tend to absorb large amounts of signals trying to pass through them. Evergreen trees are the worst because they store the most water inside them. When trouble-shooting RF, beware of any large amounts of water-based products, objects, or stock. It often occurs that someone moves large amounts of palletized cardboard boxes and RF signals in that area diminish because of absorption.

## 1.3  Signal-to-Noise Ratio

Within wireless networks, many types of interference exist. Some may be avoidable and other types are always present. The type of interference that is always present stems from the movement of electrons and the basic radiation of energy. This means that no matter what one does, there will always be a slight amount of interference present in any airspace. This small level of interference makes up what is called the "noise floor." To send a wireless signal, one must be able to transmit a signal above the noise floor. Once this is accomplished, one must overcome another interference type called "impulse noise." Impulse noise consists of irregular spikes or pulses at high amplitude in short durations. This kind of interference can be caused by a number of things, ranging from solar flares and lighting to microwaves and walkie-talkies.

The signal-to-noise ratio (SNR) helps wireless designers identify the quality of their transmissions. This is done by taking the signal power and dividing it by the noise power, producing the SNR value. This value is often measured in decibels (dB). The SNR value can help an RF designer understand how far the wireless area of coverage extends. In thinking about this, we are commonly under the mindset of increasing the power above the noise to fix our problems. Although this may be true, the FCC or, outside the United States, other government bodies regulate the amount of power a radio device can emit. However, this can impede one's ability to easily get around interference issues by increasing power. The main goal of the government's regulation is to prevent the basic radiation of energy from propagating out of proportion. If this was to happen, it would just increase the general noise floor for everyone, making it even more difficult to avoid interference.

Looking at SNR values, one needs to understand a couple of facts about different values. First, an SNR value of 3 dB is equal to 2:1, which means that the noise level is about half that of the original signal. This number doubles for every 3-dB SNR value; this means if 3 dB is 2:1, then 6 dB is 4:1. Another fact is that for every increase of 3 dB, not only does one see the noise ratio change, but one also sees that the original power level has doubled. Using surveying tools, one may find oneself losing the connection around 5 to 9 dB. This is because one is getting very close to the 2:1 noise ratio explained previously. Most surveyors use a much higher value to take in account the different power types of wireless adapters and the movement of any interfering objects, such as stock on shelves. This value strongly depends on the environment and can fluctuate from 12 to 17 dB, giving an SNR value of 20:1 on the low end and 80:1 on the high end.

## 1.4 Modulation

This section discusses some common modulation techniques so that one can get an understanding of how they work. Subsequently, this section

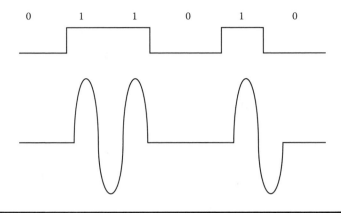

**Figure 1.5  Basic modulation.**

discusses some of the modulation techniques used by wireless networks. Before getting into the many types of modulation used on wireless networks, one must understand what modulation is and how it works to increase bandwidth on a link.

When discussing modulation, one must first focus on bits and baud and how they compare with one another. Bits, which are expressed as bit rates or typically related against time as bits per second (bps), are the measurement of data throughput. Baud is the rate of signal changes needed to send bits down a signal path. When one wants to take data and send it down a type of media such as a telephone line, it must be modulated into two different signals, which can be identified as a one (1) or zero (0). To do this, an oscillating wave is modulated by any number of techniques, such as amplitude, frequency, or phase, to create differences in the signal that can be received and returned to bits. Just like modems, wireless networks use modulation techniques to achieve communication and increase bandwidth. Looking at Figure 1.5 shows how an analog signal can be used to convey a one or zero, or vice versa.

Exploring modulation gives a good idea about how wireless networks are able to jump in bandwidth just by changing their modulation technique. It will also help us understand how wireless networks actually send information. Using modulation techniques to increase bandwidth was also seen in the rapid increase of bandwidth on modems in the late 1980s. The modem designers found better ways to modulate the data and thus increase their throughput. Before starting the modulation, one needs to make sure there is an open communication channel. A carrier signal is what is used to ensure that the communication channel is open and modulation can take place. The awful sound a modem makes is its carrier signal connecting the transmitter and the receiver together before they start modulating data.

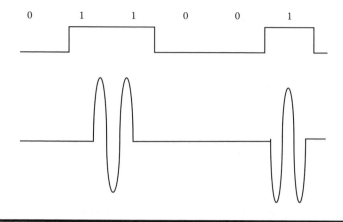

**Figure 1.6  Amplitude modulation.**

## 1.4.1 Amplitude Modulation

Amplitude modulation (Figure 1.6) is most often recognized in AM radio. This was one of the first and most basic approaches to modulation. It works by taking the signal and applying voltage to it to indicate the presence of data. When voltage is present on the line, it means a one-bit notation or "on"; and when voltage is not on the line, it indicates a zero bit notation or "off." Some coding mechanisms of amplitude modulation call out what is called a non-return to zero (NRZ); this means that if succeeding binary ones are present, the signal will continue to supply voltage for the given period of all the succeeding binary ones.

## 1.4.2 Frequency Modulation

Frequency modulation (Figure 1.7), which most people use to listen to their favorite radio stations, is another modulation technique. Another name for frequency modulation is frequency shift keying (FSK); this comes from the old telegraph system wherein the operator would key in Morse code to relay a message. To understand how frequency modulation works, let us look at the old telegraph system. When an operator was waiting for a message to be sent, the key on the telegraph system was not pressed and no signal was going down the line. Once someone wanted a message to be sent, the operator would push Morse code onto the key and each time a signal would be sent down the line. This change in frequency was from no frequency to a frequency. Once the telegraph became automatic, a signal was always present; and once each key of the message was pushed, the signal changed to a higher frequency, giving us frequency modulation.

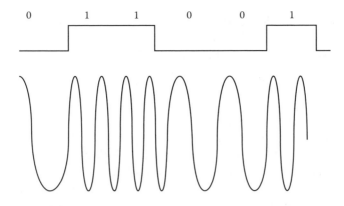

**Figure 1.7   Frequency modulation.**

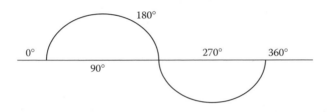

**Figure 1.8   Phase modulation.**

### *1.4.3 Phase Modulation*

Phase modulation is the one of the more common modulation techniques in use today. This is because it has the greatest ability to carry data when compared to the other modulation techniques we have looked at. Phase modulation has many different flavors itself. Some of these flavors incorporate the dual use of phase modulation and the previous techniques looked at in this chapter. A basic definition of phase modulation is the process of encoding information into a carrier wave by varying its phase in accordance with a type of input signal. Looking at Figure 1.8 provides a basic understanding of this. If one looks at a carrier wave, in this case a simple sine wave, one can see that its starting point corresponds to 0°. When the wave peaks, one has 90°; as it retunes to zero, one does not call it zero, but rather 180° because it returned from 90° and one can differentiate it from a wave just starting at 0°. In addition, one can also use the negative portion of the wave. As it reaches its negative peak, one has 270°; when it returns to zero, one has 360° instead of zero because it came from the negative peak. Now, to phase this sine wave, one needs to delay the wave's cycle. In doing this, one can see that the wave should be at 180°, when in effect it is at 270°, making it 180° out of phase.

Now that we understand phase modulation, let us see how it is used to encode data. One of the simplest ways for phase modulation to encode data is called binary phase shift keying (BPSK) modulation. In this technique, one uses a simple 0° phase change that equals a binary 0 and 180° phase change that equals a binary one. When the signal is sent without any phase changes, it represents a binary zero; when there is a change, one will see a 180° change in phase, which repents a binary one. This can be increased using the other degree markers such as the 90° marker and 270° marker. When all four phase change degree markers are used, one has what is called quadrature phase shift keying (QPSK). One can also introduce a more angular phase change; however, the more closely the phase change gets to another, the more difficult it is to distinguish the size of the signal's phase change.

In direct relation to wireless networking, there are some modulation methods to look at. The first is included in the 802.11 standard and is called differential binary phase shift keying (DBPSK). This method is similar to the binary phase shift keying (BPSK) discussed above. It uses 180° of phase change to repent a binary one and 0° of binary change to repent a binary one. This means that if the data that must be sent is 0010, the wave's signal will flow as follows. The first two zeros would be sent and no phase change would take place. Once the binary one was set to be transmitted, the phase would change to 180° out of phase. This would represent a binary one. After that, the signal would return to zero phase change, which indicates that binary zero was transmitted.

The DBPSK produced the 1-MB data rate in wireless 802.11. As we will see in Chapter 6, the 802.11 standard was capable of producing a 2-MB data rate. To achieve this, another modulation technique was used, called differential quadrature phase shift keying (DQPSK). This technique is used by a number of cellular technologies as well as the 802.11 standard. It is very much like the quadrature phase shift keying (QPSK) discussed previously. It works by having four points of reference for phase change. So, the 0, 90, 180, and 270 were used to allow encoding of more binary bits.

## 1.4.4 Complementary Code Keying (CCK)

Once the 802.11b standard was released, another modulation method called complementary code keying (CCK) was included to reach higher data rates. This method uses QPSK in a similar fashion, although it employs coding techniques to increase the coding. It is performed by a complex mathematical symbol structure that repents encoded binary bits. These symbols can endure extreme interference levels and have very little chance of being mistaken for each other.

## 1.4.5 *Quadrature Amplitude Modulation (QAM)*

When looking at modulation techniques, one sees the three discussed thus far in this section. Another method that has come out involves using two of these methods together. When one puts phase modulation and amplitude modulation together, one gets quadrature amplitude modulation (QAM). This is a technique in which both the phase and amplitude of a carrier wave are varied to allow for even more data bit encodings. In this, one not only has up, down, left, and right attributed to degrees and code bit, but also different levels of amplitude that allow for more bits to be encoded and sent.

The 802.11a and 802.11g standards outlined in Chapter 6 used a technique called orthogonal frequency division multiplexing (OFDM). Inside the OFDM standard are four types of modulation techniques: (1) binary phase shift keying (BPSK), (2) quadrature phase shift keying (QPSK), (3) 16-quadrature amplitude modulation (16-QAM), and (4) 64-quadrature amplitude modulation (64-QAM). Having discussed the first two modulation types, one can now look at the latter two: 16-QAM and 64-QAM.

Instead of using the technique discussed above, the OFDM standard took on a different approach. They used the signal constellation and broke it into four parts. Imagine an X- and Y-axis crossing to obtain a cross; inside the cross there are four distinct sections, which are used by 16-QAM to represent four subsections inside each of the original sections. This is illustrated in Figure 1.9 where one can see each of the four sections and the subsections. To change from 16-QAM to 64-QAM, one would use 6 encoded bits instead of 4 and 64 locations instead of 16. Digital television is one example of 64-QAM technology.

## 1.5 Wireless Groups

When discussing wireless groups, two main categories come to mind: (1) the wireless governing bodies on a national and international level, and (2) the bodies that create interoperability standards to promote standardization of technologies. This section outlines both of these groups and looks at how they were created, why, and what benefits they provide.

Looking closer at the first group of wireless bodies, one notices that these groups exist on a national and international level. This is because the threats of interference and the goal of creating worldwide wireless networks have always applied to each country in the world. Because of this, a global wireless body was created. Now, going back to each region or nation, one has small groups that detail the exact usage of the spectrum

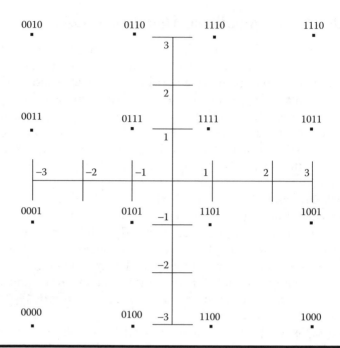

**Figure 1.9** **16-QAM.**

that is globally allocated by the international groups. Reading further into this section will provide an outline of how this works and how each of these bodies works with the other to prevent radio spectrum chaos.

The wireless industry started out with vendors designing and creating their own wireless solutions. This made each network proprietary to that vendor; and if a vendor went out of business, so did any ability to get more of the needed network equipment. Wireless groups were created to make wireless technologies better able to interoperate between multiple vendors. The creation of wireless groups led to decreased time to market for new products, as well as more interoperability between vendors. These wireless groups, or standards bodies, create the main guidelines that wireless networks must follow. These groups have many internal problems that come about between what each manufacturer thinks is right; but overall, it is much better than it was before any wireless groups existed.

### 1.5.1 International Telecommunications Union (ITU)

The International Telecommunications Union (ITU) was formed on May 17, 1865, in Paris, France. The reasoning behind this union was to streamline the process by which telegraphs were sent internationally.

Before this union, each county had expended time and resources fulfilling the requirements of each independent country. The complexity of dealing with each country and each of their requirements led to a meeting to address this issue. In this meeting, which lasted two and a half months, the ITU was created. This allowed each of the participating world governments to meet and create, agree, and modify different methods of communication. This union had 20 founding members when it was first created.

On October 15, 1947, the ITU became a specialized agency under the United Nations (UN). During this time, the ITU created the International Frequency Registration Board (IFRB) to handle the task of managing the radio-frequency spectrum. This group was in charge of the Table of Frequency Allocations, which accounted for all frequency spectrum use throughout the world.

## 1.5.2 *International Telecommunications Union Radio Sector (ITU-R)*

International Telecommunications Union Radiocommunication Sector (ITU-R) is a sub-group created by the International Telecommunications Union. The ITU-R is in charge of the technical characteristics and operational procedures of all wireless services. As part of its charter, the ITU-R develops and maintains the Radio Regulations. This regulation serves as a binding international treaty that governs the use of the radio spectrum for all of its members worldwide.

One of the key documents that the ITU-R is in charge of is the Radio Regulations. This document is a subsection of The International Frequency Registration Board's Table of Frequency Allocations. The Radio Regulations were created in 1906 in Berlin, Germany, and address the frequencies ranging from 9 kHz to 400 GHz in the Table of Frequency Allocations. Today this document contains more than 1000 pages detailing how the spectrum can be used and shared around the globe. Making changes to this document is only allowed at a world radiocommunication conference such as the World Administrative Radio Council (WARC). During this event, members discuss, create, and ratify definitions for frequency allocation.

## 1.5.3 *Federal Communications Commission (FCC)*

The Federal Communications Commission (FCC) is a United States Government agency established by the Communications Act of 1934. Its main goal is to regulate interstate and international communications. These communications include radio, television, wire, satellite, and cable. The section of the FCC that deals with wireless technologies is the Wireless Telecommunications Bureau (WTB). Its service includes cellular telephone, paging, personal communications services, public safety, and other commercial and private radio services. The WTB is also the bidding authority for spectrum auctions.

The main goals of the Federal Communications Commission's Wireless Telecommunications Bureau are to:

■ Foster competition among different services
■ Promote universal service, public safety, and service to individuals with disabilities
■ Maximize efficient use of spectrum
■ Develop a framework for analyzing market conditions for wireless services
■ Minimize regulation, where appropriate
■ Facilitate innovative service and product offerings, particularly by small businesses and new entrants
■ Serve WTB customers efficiently (including improving licensing, eliminating backlogs, disseminating information, and making staff accessible)
■ Enhance consumer outreach and protection and improve the enforcement process

## 1.5.4 Conference of European Post and Telecommunications (CEPT)

The European Conference of Postal and Telecommunications Administrations — CEPT — was established in 1959. At that time, only 19 countries were involved. As the CEPT gained momentum, it was able to expand into 26 countries within its first 10 years. After 29 years of operation, the CEPT organization decided to create the ETSI, the European Telecommunications Standards Institute. The ETSI was created to handle standardization and not regulatory issues such as spectrum allocation.

To get a better understanding of the CEPT and ETSI, one can compare them to the United States and its creation of the Federal Communications Commission (FCC) agency and the Institute of Electrical and Electronics Engineers (IEEE). The FCC manages issues concerning spectrum allocation and power regulations inside each allocated spectrum. The IEEE builds standards for devices that can operate in one of the allocated spectrums.

The members of the CEPT as of September 21, 2004, are listed as follows:

Albania, Andorra, Austria, Azerbaijan, Belarus, Belgium, Bosnia and Herzegovina, Bulgaria, Croatia, Cyprus, Czech Republic, Denmark, Estonia, Finland, France, Germany, Great Britain, Greece, Hungary, Iceland, Ireland, Italy, Latvia, Liechtenstein, Lithuania, Luxembourg, Malta, Moldova, Monaco, The Netherlands, Norway, Poland, Portugal, Romania, Russian Federation, San Marino, Serbia and Montenegro, Slovakia, Slovenia, Spain, Sweden, Switzerland, the former Yugoslav Republic of Macedonia, Turkey, Ukraine, Vatican

## 1.5.5 Wi-Fi Alliance

The Wi-Fi Alliance is a nonprofit international association formed in 1999. Its main goal is to certify the interoperability of wireless local area network (LAN) products based on the IEEE 802.11 specification. Wi-Fi stands for wireless fidelity. The Wi-Fi Alliance has certified more than 1000 products with its Wi-Fi® certification. This association came about due to the lack of well-defined technical areas in the 802.11 standard. As seen later in this book, most of the wireless standards lack certain details. For example, the 802.11 standard states that roaming will be supported, although it does not detail how a manufacturer should allow for roaming. This means that the Wi-Fi Alliance will only certify products to what is defined in the standard. The security mechanism called WEP only started as a 40-bit key in the original 802.11b standard. In the security section of this book, one sees that the key size of WEP was increased to 104 bits; this was done outside the IEEE standard. This means that for the Wi-Fi Alliance to certify a product, it only has to support a 40-bit key rather than the more often recommended 104-bit key. The Wi-Fi Alliance's goal was to make sure that if a product is Wi-Fi certified that it would interoperate with other Wi-Fi-certified products. The original name of the Wi-Fi Alliance was the Wireless Ethernet Compatibility Alliance (WECA).

## 1.5.6 IEEE

The IEEE, which is an acronym for Institute of Electrical and Electronics Engineers, is the group that created all the 802 standards. This also includes the wireless standards in the 802.11 space. The IEEE has been around since 1884 although it was not always called the IEEE. In 1963, the AIEE (American Institute of Electrical Engineers) and the IRE (Institute of Radio Engineers) merged. This came about from the existence of two separate standards bodies that were made up of many of the same people. Instead of them arranging two different meetings with each other for very similar objectives, they decided to merge the two organizations. Many brilliant minds, including Thomas Edison, were part of the AIEE, which is now known as the IEEE.

The IEEE is a governing body that created the 802 standards for network communications. The requirements needed to create an IEEE standard have a well-defined process that has seven layers. These layers allow the standard to move from thought to a written, defined, and approved IEEE standard. The seven-step process is outlined as follows:

1. Call for interest
2. Study group
3. Task group
4. Working group ballot
5. Sponsor ballot
6. Standards board approval
7. Publication

The process starts out with a call for interest in which the IEEE kicks off a meeting about a peculiar standard. In our case, this would most likely be a new wireless standard. The IEEE has a large scope, well over the small wireless subsection that relates to our example. Once the call for interest has been performed, a meeting will take place. In this meeting, attendees discuss the need for this type of standard and whether or not it is even needed. Depending on how they react to this initial meeting, the IEEE can continue with this standard or can stop it here. If they decide to continue, the next step is to develop a study group of participants to look into this further. This group would work together to discuss and decide if they are willing to commit to the next phase, in which a standard will be drafted. Once the study group moves to the task group, they are going to start writing the standard draft. Once the draft is finished, it will need to go to a working group ballot. At this point, the standard must receive a 75 percent approval rate until it can move to the next step. Most of the time, many drafts are created in this phase until one finally receives the required votes. This is when the most disagreement and time-consuming discussion takes place. Frequently, this is because each vendor involved with a particular standard has already invested R&D dollars into something that another group member may want to change. This political battle takes place until a vote reaches the 75 percent mark needed to move to the next step. After the vote meets the 75 percent mark, it goes through another ballot in which executive members of the IEEE vote on it. After this phase, it goes to the IEEE to approve and publish.

## 1.6 Chapter 1 Review Questions

1. What happens to an 802.11b wireless signal when an evergreen tree is located between the transmitter and receiver?
   a. Nothing.
   b. It is refracted.
   c. It is diffracted.
   d. It is reflected.
   e. It is absorbed.

2. What are the two correct terms used to measure antenna gain?
   a. dBi and DBd
   b. Watts and milliwatt
   c. dBd and dBi
   d. EIRP

3. The designator dBi is a decibel compared to what?
   a. Milliwatt
   b. Decibel
   c. Isotropic radiator
   d. RADIUS

4. What does RF stand for?

5. Which type of modulation does 802.11b use?
   a. QAM
   b. FM
   c. AM
   d. CCK

6. How can one send more data across the air?
   a. Increase the transmit power.
   b. Use a more complex modulation.
   c. Use a bigger antenna.
   d. Use a wider frequency band.

7. What was the Wi-Fi Alliance formerly known as?
   a. FCC
   b. WECA
   c. IEEE
   d. WIFI

8. What seal certifies interoperability in a manufacturer's wireless device?
   a. WHY
   b. Hi-Fi
   c. WECA
   d. Wi-Fi Certified™
   e. Wi-Ki

9. Which wave will travel the greatest distance?
   a. FM radio
   b. X-ray
   c. 802.11a
   d. Microwave

10. What two items should be maintained near the edges of a wireless cell when performing a site survey?
    a. High signal-to-noise ratio
    b. Low signal-to-noise ratio
    c. High noise level
    d. High signal strength

11. What would the FCC and ETSI regulate on a wireless network? (Select more than one)
    a. Power outputs
    b. Total client number
    c. Channel number and frequency
    d. Who can use the system

12. What bandwidth term is this phrase stating? On any given day, my wireless network has a low bandwidth of _____.
    a. 11 Mbps
    b. 2.4 GHz
    c. 11 MHz
    d. 5.4 GHz

13. Which of the following show the correct use of a wireless network?
    a. Using wireless to connect two buildings point-to-point
    b. Mobile access from laptop or PDA
    c. As a way to connect a server
    d. To increase bandwidth on a 10/100 wired network

14. When a wireless signal changes or bends around an object, sometimes creating a shadow area, it is known as _____.
    a. Refraction
    b. Reflection
    c. Diffraction
    d. Scattering

15. When performing wireless power calculations, what two terms are often converted into each other?
    a. dBi to dBi
    b. dB to Watts
    c. dBm to DBi
    d. dBd to dBi

16. What standards body creates wireless standards?
    a. Wi-Fi Alliance
    b. IEEE
    c. FCC
    d. WECA

17. What key size is required in WEP for the Wi-Fi Alliance to certify a product?
    a. 40 bits
    b. 120 bits
    c. 128 bits
    d. 56 bits

18. What does Wi-Fi stand for?
    a. Wireless infrastructure fidelity industry
    b. Wireless Interoperability Forum Institute
    c. Wireless fidelity
    d. Wireless networking

# Chapter 2

## Risks and Threats of Wireless

This chapter discusses the general goals for information security and how they are used to measure risk and understand threats. This information will help in the next sections of this chapter where each of the threats relating to the many types of wireless communications is explored. After looking at each of the threats, this chapter focuses attention on wireless hackers. In this chapter, we see how hackers locate the existence of wireless networks as well as how law enforcement tracks down these hackers.

## 2.1 Goals of Information Security

When looking at information security, one must address the three tenets of information security: (1) confidentiality, (2) availability, and (3) integrity. These long-standing goals will help us understand what we are trying to protect and why. This information will help when one starts looking at all the risks and threats that face wireless communications. Before one can properly evaluate risk, one needs to set a baseline to understand the definition of each goal one is trying to uphold.

## 2.1.1 Confidentiality

Attacks on the confidentiality of information relate to the theft or unauthorized viewing of data. This can happen in many ways, such as the interception of data while in transit or simply the theft of equipment on which the data might reside. The goal of compromising confidentiality is to obtain proprietary information, user credentials, trade secrets, financial or healthcare records, or any other type of sensitive information.

Attacks on the confidentiality of wireless transmissions are created by the simple act of analyzing a signal traveling through the air. All wireless signals traveling through the air are susceptible to analysis. This means there is no way to have total confidentially because one can still see a signal and subsequently record it. The use of encryption can help reduce this risk to an acceptable level. The use of encryption has its own flaws, as seen later in this book. For the most part, the encryption is secure itself, although how it is implemented and how key management is handled may produce flaws that are easily exploited.

## 2.1.2 Availability

Availability is allowing legitimate users access to confidential information after they have been properly authenticated. When availability is compromised, the access is denied for legitimate users because of malicious activity such as the denial-of-service (DoS) attack.

Receiving RF signals is not always possible, especially if someone does not want you to. Using a signal jammer to jam an RF signal is a huge problem that has been facing national governments for years. Looking for the availability of RF local area networks (LANs), one notices that performing a DoS attack is easy to accomplish. This is due to the low transmit power allocated by the U.S. Government and poor frame management techniques included in most of the current day wireless standards.

## 2.1.3 Integrity

Integrity involves the unauthorized modification of information. This could mean modifying information while in transit or while being stored electronically or via some type of media. To protect the integrity of information, one must employ a validation technique. This technique can be in the form of a checksum, an integrity check, or a digital signature.

Wireless networks are intended to function in an unimpaired manner, free from deliberate or inadvertent manipulation of the system. If integrity is not upheld, it would be possible for an attacker to substitute fake data.

This could trick the receiving party into thinking that a confidential exchange of data is taking place when in fact it is the exact opposite. Wireless networks have adapted to this type of threat over time. One can see this advancement as new security standards emerge, creating increasingly complex integrity checks.

## 2.2 Analysis

Analysis is the viewing, recording, or eavesdropping of a signal that is not intended for the party who is performing the analysis. All RF signals are prone to eavesdropping; this is because the signal travels across the air. This means anyone within the signal's path can hear the signal. One of the only protections available to prevent the loss of confidentiality is encryption. If a signal is using encryption, then its confidentiality can be upheld until that form of encryption is defeated. The risk of analysis on an RF signal is an inherent risk that cannot be avoided. The only option is to mitigate the risk with some type of confidentiality control.

## 2.3 Spoofing

Spoofing is the act of impersonating an authorized client, device, or user to gain access to a resource that is protected by some form of authentication or authorization. When spoofing occurs in wireless networks, it primarily involves an attacker setting up a fake access point to get a valid client to pass authentication information to that attacker. Another way attackers spoof is by performing a man-in-the-middle attack. In this scenario, an attacker would position himself between a client and the network. This could be accomplished by spoofing a valid access point or by hijacking a session. Once this part is complete, the attacker would then use the authentication information provided by the client and forward it to the network as if it originally came from the attacker.

## 2.4 Denial-of-Service

Denial-of-service (DoS) is the effect of an attack that renders a network device or entire network unable to communicate. Hackers have found that certain crafted packets will make a network device unresponsive, reboot, or lock up. They have used this technique to shut down high-traffic networks and Web sites. They have also used this attack to reboot network equipment in an attempt to pass traffic through the device as it is booting up. This is done to try to circumvent any policies set up on the device to protect it or devices behind it. The DoS threat can also adversely affect the availability of a network or network device.

Wireless DoS attacks can be achieved with small signal jammers. Finding signal jammers is not as difficult as one might think. Some modern-day wireless test equipment can perform jamming. This is not the tool's intended purpose, although it is commonly used for this. Jamming is possible because the government regulates the amount of power allowed on a wireless network. In relation to wireless LANs, the amount of power used is a very small amount. This means that it is not difficult to overpower an existing device with a home-made one.

Another DoS threat relating to LANs in particular is the poor structure of management frames. These frames allow for anyone who can analyze the wireless signals to perform a DoS attack by replaying certain management frames. Mostly, theses attacks are layer two frame attacks. These attacks try to spoof management traffic, informing the client that he is no longer allowed to stay connected to the network. Chapter 13 discusses these attacks in more detail.

## 2.5 Malicious Code

Malicious code can infect and corrupt network devices. Malicious code comes in many forms: viruses, worms, and Trojan horses. People often confuse the three main forms of malicious code. Because of this, they use these terms interchangeably. This section looks at each of these and identifies what classifies them into each of the three groups. Viruses infect devices and do not have the ability to replicate or spread outside the infected system on their own. Once a virus infects a machine, it can only replicate inside the infected machine. This means that all threats from viruses stem from receiving infection. The threat of worms is much higher because they can spread across the enterprise and out to the Internet, infecting multiple devices. In the past few years, humans have started to see global worms that propagate across the entire world. The final malicious code threat discussed here is the Trojan horse. This threat comes from installing or running programs that can have or within their use execute code that might contain malicious content.

Malicious code relating to wireless has to do with new viruses that can affect the many new types of wireless end devices such as PDA units, smart phones, PDA phones, laptops, etc. Wireless viruses have just started to appear in the wild. Even with this threat just starting to develop, many forms of wireless malicious code have already appeared. Some of this code has enough intelligence to find and utilize a variety of available wireless technologies on a device to spread even further.

Another form of malicious content relating to wireless is *spam*. Although spam is not destructive in nature, the time and money it costs an organization often makes it seen as malicious. Spam is not just related to wireless. Long before wireless spam there was e-mail spam. Today's wireless devices are capable of receiving messages in many formats: e-mail, text messaging, instant messaging, and voice calls. All of these are starting to see spam pop up on them. Dealing with spam has created a security market of its own with products, solutions, and services created to combat this threat.

## 2.6 Social Engineering

Social engineering is the often called low-tech hacking. It involves someone using the weakness of humans and corporate policies to obtain access to resources. Social engineering is best defined as tricking or manipulating a person into thinking the party on the phone is allowed access to information, which they are not. The threat of social engineering has been around for quite some time. Some of the most well-known computer hackers used this type of attack to get information. The real threat to this is the skill level involved. No one needs to be computer savvy or a technical genius to perform this type of attack. There are a number of things to do to prevent this type of attack. First, make sure that a policy is in place regarding sensitive information and phone usage. Make sure that not anyone can call and reset someone's password. Create a help-desk identification process to authenticate callers to the help-desk operators.

## 2.7 Rogue Access Points

Rogue access points pose a major threat to any organization. This is because of the high availability and the limited security features of off-the-shelf access points. If a company does not approach the WLAN (wireless local area network) concept fast enough, frustrated employees will take it upon themselves to start the process. When this happens, employees often put in wireless systems of their own. Even with most current-day access points supporting advanced security standards, the default configuration of an out-of-the-box access point is set to the least secure method. This has created a real threat because now a user can easily bring in a rogue access point, plug it in, and put the entire network at risk. The knowledge level required to install an off-the-shelf access point has almost become plug-and-play today. This means that more and more people have the ability to place rogue access points. These same people lack the ability to secure these devices or even understand the risk they are posing for the company.

Most access points come from employees, although as we will learn later there are cases where an attacker would try to set one up for easy return access. This was not a big issue until recently when the price of 802.11b access points fell well below $100. To do this, an attacker would need physical access and a network port. If a hacker wanted access bad enough, spending $100 for it would be a conceivable expense.

With companies investing in stronger security mechanisms, it would be a shame to have an incident in which an attacker gains access through a non-secure rogue access point. Because of the threats associated with rogue access points many companies have started to put controls in place to increase awareness and prevent the deployment of rogue access points. Many companies that jumped into the newly formed wireless security market have adapted and created tools to detect rogue access points. Some companies have handled rogue access points by creating policies about wireless usage and strict penalties for rogue access placement. Others have taken a second route and invested in wireless intrusion detection systems (WIDS) software.

## 2.8 Cell Phone Security

Now we will have a discussion of general cell phone identification and security. Cell phones have had a slight advantage over other types of wireless communications in the security realm. This is due to their over-whelming numbers. Most people today have a cell phone; and with so many people using cell phones, many security risks and subsequent controls have been developed to counter each other. Understanding this information will show how cellular phone providers have mitigated similar risks that face wireless local area networks.

Cell phones send radio frequency (RF) transmissions on two distinct channels: (1) one for actual voice communication and (2) the other for control signals. This control signal identifies itself to a cell site by broadcasting its mobile identification number (MIN) and electronic serial number (ESN). When the cell tower receives the MIN and ESN, it determines if the requester is a legitimate user by comparing the two numbers to a cellular provider's subscription database. Once the cellular provider has acknowledged that the MIN and ESN belong to one of its customers, it sends a control signal to permit the subscriber to place calls.

Like all RF devices, cell phones are vulnerable to eavesdropping and spoofing. In the cellular phone industry, these are called "call monitoring" and "cell phone cloning." Another risk associated with cell phones is the ability to reprogram phones, transforming them into advanced microphones capable of recording and transmitting sound from their location to anywhere in the world.

Monitoring calls is an easy task, especially for phones that use analog technology. This is because most analog cell phone technologies were transmitted in the same band as FM radio. A commonly available radio frequency scanner could get one up and listening to calls in minutes. With the proliferation of digital cellular networks, more and more security was erected. This was great because inside a service provider's network, your calls were, for the most part, safe. There were easier analog targets for criminals to exploit. One's digital phone was not so safe if one roamed or went outside of a provider's area of coverage. When two cellular providers wanted to hand off calls to each other for billing purposes, they converted them to analog so they had a common protocol for interoperability. This also meant that security was no longer present. So, even with a digital phone, once the MIN and ESN are removed or identified from the phone call, it could still be tracked, cloned, or monitored inside the digital network.

Another trick involves turning a cellular telephone into a microphone and transmitter. This can be used to record a conversation or bug a room. This can be done without your knowledge by police, governments, and even some highly educated people. How does it work? It is easy to do, just send a maintenance command on the control channel to the phone. This command places the cellular telephone in a diagnostic mode. When this is done, conversations in the immediate area of the telephone can be monitored over the voice channel. The signal engages the phone to perform this monitoring action without any indication of it taking place. The user does not know the telephone is in the diagnostic mode and transmitting all nearby sounds until he or she tries to place a call. The calling feature does not work and the phone is useless until the power is cycled. After that, the phone returns to a normal state as if nothing ever happened.

This is very scary because the user has no idea he is bugged by his own phone through the airwaves. This threat is the reason why cellular telephones are often prohibited where classified or sensitive discussions are taking place. Someone could be bugging your phone as you read. Do not worry; as long as one can place a call without cycling the power, you're ok.

One publicized case of cell phone monitoring involved former Speaker of the House of Representatives, Newt Gingrich. A call between Gingrich and other Republican leaders was monitored and taped. The conversation concerned Republican strategies for responding to an ethics violation for which Gingrich was being investigated. This call was given, or most likely sold, to the *New York Times* and made public.

Another publicized case of cell phone monitoring involved a pager system instead of a cellular phone system. In 1997, the Breaking News Network monitored the pager messages of a large number of New York City leaders, including police, fire, and court officials. The messages recorded were considered too sensitive to send over the government's protected police radio. This confidential information was captured and then sold to other news agencies in order to get the scoop on a story. This ended up happening sometimes before the police dispatch even had the information. Later in the year, police arrested the officers of this New Jersey news company for illegally monitoring their pager systems.

Next we look at cellular phone cloning. What is cell phone cloning? It is the copying of the unique identification information programmed into your cell phone by a cellular provider. The cellular provider programs the phone with an electronic serial number (ESN) and mobile identification number (MIN). A cloner will steal this information, copy it to a different phone, and place calls on your bill.

There are many ways for cloners to obtain these numbers. One is when someone fixes your phone or even when you buy a new phone at the store, someone could copy this information during the activation process. The MIN and ESN can also be obtained by an ESN reader (see Figure 2.1), which is similar to a cellular telephone receiver designed to monitor control channels. The ESN reader captures the MIN and ESN as they are being broadcast from a cellular telephone to a cell tower. This happens when your phone is turned on or when you move from one cell phone tower to another.

A major controversy grew around cell phone cloning. At first, the phone companies refused to admit that their security was compromised, thus making the victim pay for all the calls placed by the cloner. This proved to be a big problem for cell phone companies and their customers.

Another threat related to cellular phones deals with the short messaging service (SMS). This is a method of sending short messages similar to e-mail. One of the threats related to this has to do with mass SMS messages that create a denial-of-service attack. This sort of attack has not been widely seen yet, although many industry leaders have openly spoken about the risk and impact if it were to happen.

## 2.9 Wireless Hacking and Hackers

Inherently, RF has many threats, including interception, signal jamming, and signal spoofing. Because RF travels through the air, picking up the signal is as easy as being within the radio waves' vicinity with the right hardware. Spectrum analyzers can detect radio transmissions showing the

**Figure 2.1   ESN Reader.**

user the signal frequency. Depending on that frequency, an attacker might be able to identify the transmission right away. Most RF frequencies in the spectrum are reserved for specific uses. Once one is able to find a signal and map it to a reserved spectrum, one knows who is transmitting it and, in some cases, why.

Getting more in tune with the majority of RF threats, one can look at today's RF local area networks (LANs). These, of course, have the same threats as all RF signals, although they do add a new dimension stemming from mass use and scrutiny. Just like cell phones, the more people who use them, the more time people spend looking at how they work and what they can do to defeat any security that exists. This has been seen over the past few years as a large number of users started deploying wireless networks and security flaws began to pop up. Most wireless network setups are capable of working right out of the box. This has led more and more nontechnical people to deploy them. When setting up a wireless LAN right out of the box, the default configurations are usually the most insecure ones.

## 2.9.1 Motives of Wireless Hackers

America's laws and law enforcement agencies have taught us that rarely is a crime committed without a motive. With this said, if someone was to spend the time to compromise an RF signal, there always is a motive. Some of these motives can be as harmless as wanting an Internet connection to send a loved one an e-mail, or as terrible as committing an act of terrorism against a nation or government. To understand why someone would try to compromise an RF signal, take a look at some of the more well-known motives, such as to get a free Internet connection, commit fraud, steal sensitive information, perform industrial or foreign espionage, and — the worst of all — terrorism. After understanding what motive or motives an attacker might have, one can better understand how much security one should apply to the RF signal. If a company deals with financial information, it probably is more at risk from an attacker than a small doll shop. Knowing who might attack and why can help ensure that the correct risk-reducing actions are taken.

## 2.9.2 War Drivers

After more and more people realized that out-of-the-box wireless LANs were generally set up, by default, in the most insecure mode, people started to exploit them. This new fad of identifying and categorizing wireless networks based on their security level has been coined "war driving" (see Figure 2.2). War drivers use equipment and software to identify wireless networks. War driving allows attackers to understand the security associated with any particular wireless network they happen to pass. This equipment not only allows war drivers to identify, locate, and categorize wireless networks, but also allows them to upload their results

**Figure 2.2  War drivers.**

to a Web site where their friends and everyone else who has access will be able to see exactly where these unsecured wireless networks are located. This has even become so advanced that the use of GPS has been incorporated to give other people exact locations to these insecure networks. Anyone can simply go online and get a map of the exact location of an insecure network identified by a war driver.

This culture has taken on its own members who are not stereotypical classic teenage computer crackers or hackers. Some older, wealthier people have begun to war drive. One individual turned his Acura in-dash GPS into a war-driving display. Others have mounted fixed wireless antennas onto their vehicles. Figure 2.2 shows a war driver using a Cadillac.

Today, there are major events as well as Web sites dedicated to indexing and storing wireless network location information. They use information provided by war drivers, and some have gone as far as organizing worldwide war drives. The Web site www.worldwidewardrive.org has done just this for four years running. Their first worldwide war drive started on August 31 and ended on September 7, 2002. In this short period of time, 9374 access points were found; and of those, only 30 percent had any encryption technology. As time passed, the worldwide war drive went into round two, lasting from October 26 to November 2, 2002. In this round, they found 24,958 access points, with the number of unsecured access points rising 2.2 percent. With rounds three and four, the number of unsecured access points fell, although they only fell a small percentage. With the fourth worldwide war drive now complete, the total number of wireless networks running without any type of encryption is 61.6 percent. Figure 2.3 has a chart that shows the data collected from all four worldwide war drives. This information was obtained from the worldwide war drive Web site. The latest information available about this year's or upcoming years' worldwide war drives is also available on that site.

## 2.9.3 War Walkers

A more athletic approach involves war walking instead of war driving. In this concept, a war walker would stroll down the street with a laptop either in a bag or out in the open. They would use the same tools and equipment employed by the war driver to identify insecure networks. This has gained a lot of steam, given that the number-one profile for a computer hacker is a teenager who most likely does not have a car or is not old enough to drive. Consumer industries have even gotten on this bandwagon by producing tools to find wireless networks. There are even devices that connect to your keychain that will beep or light up when wireless networks are detected. This has turned the act of war walking into an event that can be performed without any effort during any activity that requires

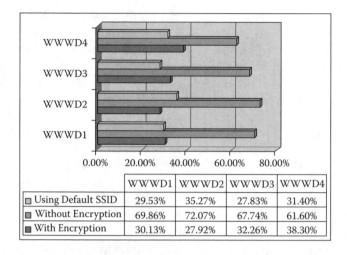

| | WWWD1 | WWWD2 | WWWD3 | WWWD4 |
|---|---|---|---|---|
| Using Default SSID | 29.53% | 35.27% | 27.83% | 31.40% |
| Without Encryption | 69.86% | 72.07% | 67.74% | 61.60% |
| With Encryption | 30.13% | 27.92% | 32.26% | 38.30% |

**Figure 2.3  Worldwide war drive stats.**
*Source:* www.worldwidewardrive.org.

movement. This means that a war walker can perform this malicious activity while he or she gets milk for mom.

## 2.9.4 War Chalking

Next is the concept of war chalking (Figure 2.4), in which a war driver or walker does not have the time to go to a Web site and locate an insecure network or does not have any available internet connection to do so. To help a fellow war driver or walker, the community has developed a number of symbols representing wireless networks and their associated security levels. This helps them to find the quickest, most insecure networks so that they can connect to the Internet and anonymously surf the Web or participate in any other activity on the Internet in an anonymous manner.

## 2.9.5 War Flying

Interestingly enough, lately a new breed of identifying wireless networks has emerged, called war flying. As the name infers, war flying is the act of scanning wireless networks from inside the cockpit of a private plane. When a person is war flying, he can cover a large distance quickly. This has proven the quickest way to collect data about insecure wireless networks thus far. This is because of the use of private planes that can fly at a low altitude and cover a large distance rather quickly. The recommended altitude is less than 2500 feet for optimal wireless scanning.

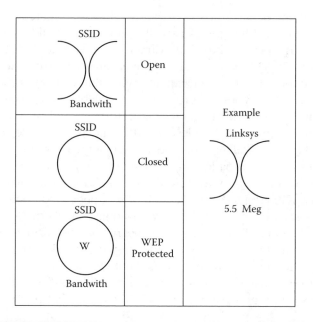

**Figure 2.4  War chalking.**

With war flying, we can see that this new trend of wireless scanning also appeals to people wealthy enough to purchase or rent a private plane.

## 2.9.6 Bluejacking

Bluejacking is a relatively new term that focuses on Bluetooth-enabled devices. Unlike the name, bluejacking is not stealing or hijacking Bluetooth devices: rather, it is a way to send anonymous messages to Bluetooth-enabled devices. A bluejacker will go to a place where there are a large number of people, such as on a subway, and scan the airwaves for Bluetooth devices. By default, many Bluetooth devices allow for pairing. When a bluejacker finds one of these devices, he sends messages to them and tries to identify them in the crowd. Once identified, messages that are more personal can be sent. This will generally make some nontechnical people a little scared. Could you imagine getting pop-up messages on your cell phone describing where you are and what you are wearing?

## 2.9.7 X10 Driving

Lately, many X10 cameras are popping up in more than just your browser. With an inexpensive wireless camera out on the market, many people are using them for all kinds of reasons. This has created what is now known as camera driving. Just as a Peeping Tom would look into your windows, these people buy X10 camera receivers and watch what is broadcasting on X10 cameras in any given area. Because these cameras have no security, any receiver can see what any camera is transmitting.

There are other types of X10 devices available to automate a home. These devices can open the garage door, turn on and off the lights, and control some appliances. Similar to X10 cameras, these devices can be controlled by any receiver. This means not only can someone see what the X10 cameras are broadcasting, but they may also be able to get into one's home or create problems with one's lighting or appliances.

Recently, X10 adopted a weak form of security called home and device codes. This security is applied by two 16-digit wheels that can be set to a house code and a device code. This means that only 256 possible codes exist. Although this might hinder the casual eavesdropper from easily picking up the devices, anyone who wants to take the half an hour to switch the code wheel between all 256 possible codes will be able to defeat the security and control any X10 device in that home.

## 2.9.8 Cordless Phone Driving

When it comes to phone driving, most of this type of eavesdropping has been reduced by U.S. Government laws and enhanced security. Older 900-MHz phones were primarily analog and could easily be picked up by most scanners. With the advent of 2.4-GHz and 5.8-GHz cordless phones, many higher frequency scanners were more difficult to procure. In an attempt to preserve privacy, the U.S. Government made a large number of these scanners illegal to own or operate. This did not stop resourceful people from tapping into phone conversations, so phone manufacturers employed a security mechanism called Digital Spread Spectrum (DSS), which sends the call information in a digital format across multiple channels.

## 2.9.9 War Dialing

We have discussed terms such as "war driving," "war chalking," and "war walking." These new hacking terms originated from an old term called "war dialing." These war dialers would dial multiple numbers looking for a modem connected to a computer system. This was because of the large number of insecure modems connected to many computers and computer networks. Applying this to wireless, there are now war drivers who drive around looking for wireless access points connected to networks. War dialing and war driving are similar in the sense that the attacker is trying to connect to a predominantly insecure medium in hopes that the sheer number of deployments will result in poor security.

## 2.9.10 Tracking War Drivers

After confronting a war driver once, I asked him, "Why do you do it?" His statement to me was rather interesting in the fact that it was a contradiction of terms, yet so many of his peers also had the same answer. One response was "to educate the public to the existence of these insecure means of accessing networks and the Internet." The second response was "amenity or the ability to go online without any record of it being traceable back to them." Now, thinking about this, if the goal was to have an untraceable Internet connection, then why expose the networks to the public eye? Well, at least the question was answered like a true politician. Credit must be given to a teenage hacker who has professional speaking skills like that.

How would someone track down a war driver? The FBI had several public cases of arresting criminals using wireless networks to compromise retail store networks. They were tracked down so we know it is possible; so let us learn how.

Once the investigation starts, a forensic team arrives on site and dumps the configuration and stored memory of all network devices and servers that were affected. Once this data has been properly removed, in accordance with the chain of evidence, it is properly examined at a lab. This examination process is a timely one, so much so that it can make many incidents considered not financially worth the effort. Many cases are too small to warrant the massive effort needed to investigate.

After the lab results are examined, one can see where the perpetrator first entered the network. Because this was on a switch connected to an access point, one can determine that they came in over the airwaves. Once this information is identified, one can determine the wireless network interface card's MAC address. This address is hard-coded onto the card by the vendor and is regulated in a sense, which makes it globally unique. Some clever hackers have the ability to change the card's MAC address, but as time has shown, many do not take the time to do this. Chapter 13 ("Breaking Wireless Security") discusses how this is done and what tools are out there to perform this type of attack.

After the MAC address has been determined, one of two things can happen. First, the police can get a warrant to search any suspect's home for the network card in hopes of finding it and its matching MAC address. In a highly important case, such as one that involves terrorism, the FBI might go back to the card maker and track that card's movement from creation at the manufacturer's factory, to the distributor, then to the retail store and finally to the purchaser. This is an easy task to accomplish, although it is very time consuming. It works by correlating many different data sources to limit the number of people to question. Looking at the tracking of the card itself, a vendor can show proof of its arrival at a warehouse or retail store. Once it is proven that it has arrived at a retail store, one only needs to find out who bought it. The first round would be to look up all transactions on the point of sales machines for anyone who purchased any of the vendor's network cards. Looking at this gives credit card information for anyone who used that method of payment. Most likely, if someone were going to do something illegal, he or she would have paid in cash. Well, it is also easy just to look on the store's video camera correlating all the times that any of the vendor's cards were purchased minus any purchases made with a credit card.

After looking at how wireless war drivers can be tracked, one gets to a more important point about wireless devices. All bi-directional communicating wireless devices emit radio waves; so in a sense, all wireless devices can be tracked in one form or another. As one reads through this book, one will see that most modern-day wireless devices have some type of tracking method associated with them. Next time you see some amazing new RF technology, remember the statement above. No matter what manufacturers say about their technologies, any **bi-directional** communicating wireless device can be tracked.

## 2.10 RFID

The Radio Frequency Identification (RFID) concept has created some major privacy issues. With RFID, companies can save time and money by being able to track products from their creation, to their purchase at a retail store, and beyond. It is the "beyond" part that has so many people upset about the inherent piracy issues of RFID. Before delving into those concerns, one needs an understanding of the technology. RFID systems have been used for quite some time. Only recently has their true potential been realized. An RFID system is a small tag that is affixed to an object to allow that object to be tracked. Once this tag has been turned on or energized, it will send information about itself when a reader queries it. This tracking can take place wherever there is a reader ready to query the tag. This means other companies can read RFID tags from their suppliers. There also is the ability to add to these tags; if one company buys a product from another and wants to insert some of its own data into the tag, they can. Another RFID innovation that has been discussed prevents something almost all of us likely have done. How many people have ended up with a pink load of white laundry? Some washing machine manufacturers have talked about using RFID tags embedded inside clothing to prevent the red sock from getting into the load of white laundry. How about never reading tags for washing items again? The same manufacturers have talked about having washing machines set themselves based on the clothing item's RFID tag.

Most RFID systems today have a write-once tag, which means that erasing or modifying RFID information is unlikely. When the price of modifiable RFID tags come down in price, a new integrity threat will emerge, called RFID modification. Today, with RFID, companies can track their products and amass amazing amounts of data. There are so many ways to use RFID. Retail companies use RFID to perform automated inventory. Car manufacturers use them to tag special-order vehicles. Logistics companies use them to track package movement. There are even more usages that are created everyday; for an example of just how massive the push for RFID is, think about this: some companies have started to look at taking data from the RFID to feed financial reports, so investors know at any given time how many units were sold or shipped per quarter. This information could have a direct result on the price of stock. This could then, in turn, affect the way stock trading is performed in the future. Just imagine if a stockbroker could see in real-time the number of units a company is selling. Now that we have a good idea of how RFID works, let us look at the inherent risks and threats involved with its use.

When discussing RFID, the first thing that comes to mind is the concern over privacy. In a world where the products one consumes transfer information to anyone willing to listen, the opportunity to market, trend, and collect data about us becomes a real concern. Some people have talked about many things relating to RFID, from the wild conspiracy theories to real issues that affect everyone on an every-day basis. To understand these concerns, one can look at a couple of examples that range from wild conspiracy theories to those that affect almost everyone every day.

Using RFID, people with access to the right information assets can track individuals based on what they have purchased. If someone buys a can of soda from one store and then walks into another store, the second store's readers might pick up that can. If someone wanted to find a person and had the resources, he or she could find the RFID tag ID number and cross-reference that with a credit card system or company database. This would allow a simple object to become a tracking device. This is highly inconceivable today; although inconceivable or not today, it is technically possible.

Another privacy concern that affects almost everyone is the ability to read and use information from product tags not belonging to the reader's organization. To put this into context, imagine walking into a store with a bag or purse. When you walk in, card readers at the door energize all the items inside your bag. Then these items send all their information to the reader. The store now has a record of your purchases, not from that store but just in general. Everything inside your bag that has a UPC would have an RFID. These records could include over-the-counter medications, feminine products, and a number of other things that many people consider private. To make things worse, the salesperson might be given this information to get an idea about your purchasing habits.

## 2.11  Chapter 2 Review Questions

1. Which of the following processes will not lower the risk of social engineering at a help desk?
   a. Positively identifying the caller
   b. Using a callback method
   c. Shredding documents
   d. Having the caller verify the identity of the help desk operator

2. What would a hacker whose motive was money most prefer to attack?
   a. School
   b. Bank
   c. Doll shop
   d. Pizza shop

3. Which of the following terms best describes a Wi-Fi hacker?
   a. War dialer
   b. Hacker
   c. War hacker
   d. War driver

4. What type of malicious code infects devices and does not have the ability to replicate or spread outside the infected system on its own?
   a. Worm
   b. Virus
   c. Trojan horse
   d. Spam

5. List the three main goals of information security.
   a. Integrity
   b. Encryption
   c. Availability
   d. Confidentiality
   e. Scalability
   f. Protecting

6. What two pieces of information are required to hack a cell phone?
   a. MNN and ESS
   b. MIN and ESN
   c. ENS and MSN
   d. Phone ID and vendor ID

7. What piece of information is unique on every wireless card in the world?
   a. IP address
   b. Serial number
   c. MAC address
   d. SSID

8. Which of the following terms are used to describe the process of discovering wireless networks?
   a. War flying
   b. War walking
   c. War driving
   d. All of the above

9. A self-replicating and often self-sending piece of malicious code, which is often e-mailed, is called _____.
   a. A worm
   b. A virus
   c. A Trojan
   d. Spam

10. What technique would an attacker do to force a wireless end device to disconnect from the network?
    a. Wireless scan
    b. Port scan
    c. OS fingerprinting
    d. RF jamming

11. What would a wireless hacker in his early teens most likely be doing?
    a. War flying
    b. War driving
    c. War walking
    d. War gaming

12. What does the term "spam" mean?
    a. Ham in a can
    b. Sending wanted e-mails
    c. Sending unwanted messages
    d. Sending junk snail mail

*Chapter 3*

# The Legality
# of Computer Crime

Throughout time, legislation has often been unable to keep up with technology growth. This is because of the rapid development of technology and the slow process of creating laws. It is important to understand what exactly constitutes a crime when any information security matter is involved.

When discussing the legality of computer crime with respect to RF transmissions, a number of issues often arise. For example, in some states it is illegal to connect to an unauthorized network. With technology providing as much usability as possible, when does one cross the line? New wireless laptops are capable of connecting to insecure networks on their own. This means that merely walking down the street can be considered a crime. For example, the FCC can prosecute a person for eavesdropping, just for enabling a wireless sniffer and capturing packets from an unauthorized network.

Think about this: when troubleshooting a wireless network in an office complex, a person might end up firing up wireless sniffers and finding out that the office suite next door has a wireless network. That person has just sniffed the airwaves and committed a crime subject to the FCC ruling. These examples reveal how different types of technology and legislation often are not finely tuned to each other, thus creating large gaps that allow criminals to exploit flaws and honest people to be criminalized by the same ones. One does not have to worry that using a wireless sniffer will get one sent to jail. For example, it is illegal for people in Michigan to spit on the sidewalk. However, no one has been convicted of this crime in the past 50 years, although it is still listed as a law.

This chapter shows the relevant legal matters affecting wireless networks. Some of the laws that were written for computer communications are not specific to wireless communications. Either way, most laws detail what constitutes a crime in relation to computers and network systems. The current chapter looks at the laws and acts that have created a good definition of what computer hacking is. Having a good understanding of that, this chapter goes on to discuss what crimes fall under these acts and laws. Just because a law says that breaking into a computer system is illegal, a process must be put in place to prevent every small computer crime from slowing down the court system. To fix this, certain requirements have been created to identify what types of crime are subject to these laws and acts, thus receiving federal aid and investigation.

## 3.1 Electronic Communications Privacy Act

The Electronic Communications Privacy Act was enacted in 1986. This was to combat disclosure and interception. The main purpose is to protect the privacy of electronic communications. This law protects U.S. citizens from any inappropriate use of any intercepted electronic transmission. Electronic transmissions are defined by this act as "the transfers of signs, signals, writing, images, sound, data, or intelligence of any nature transmitted in whole or in part by wire, radio, electromagnetic, photo-electronic, or photo-optical systems that affect interstate or foreign commerce." This act also requires government agencies to follow the correct procedure to obtain electronic communications from service providers. What this means to us has both positive and negative implications. The government must use due process and get a court order to tap any communications over an electronic device (e.g., a phone or e-mail).

On the negative side of the act, it can allow an employer to listen and record an employee's phone conversations, e-mail messages, and any other type of electronic communication taking place on the employer's equipment. The last part of this statement is the most important. The employer must own and provide to the employee the electronic communication device. If the employer provides this device to the employee, then the employer can listen to the employee's communications for a number of actions. Another key point is how and what employers can do with that information. The law has shown that if an employer wants to use any of the recorded communication for action against the employee, then the employer must first notify the employee that his or her communications are being intercepted or recorded. An even more interesting point relates to free e-mail services on the Internet. These services can use a person's e-mail for a magnitude of purposes. They can sell your likes to related retailers, and they can market products directly to you based on your purchases. The main reasoning behind this is that the law states that the information residing on a service provider system is owned by the service provider even if it is your data. The bottom line to this statement is to be careful with any outsourced data endeavor; although contracts and privacy policies can protect you and your data, the law without these added documents does not provide any recourse from the misuse of data if it, in fact, resides on someone else's system.

## 3.2 Computer Fraud and Abuse Act

Enacted in 1996, the Computer Fraud and Abuse Act helped law enforcement fight the many computer crimes that were difficult to prosecute. This was because technology very rarely keeps up with legislation. This often creates many gaps that criminals exploit. This act allowed for the identification of what exactly constituted a federal computer crime in order to close these gaps. With a clear definition of what a computer crime is, this act has helped law enforcement agencies convict computer criminals who have evaded them in court thus far.

The focus of this act is not to protect users, but rather to protect the federal government and its computer systems from any threat to national security. During the late 1980s and early 1990s, hackers were infiltrating government-controlled systems frequently. Without this federal law, prosecutors had trouble convicting these hackers of actually committing a crime. This is because of the lack of legislation dealing directly with computer crimes and exactly what constitutes them as crimes. Now with the Computer Fraud and Abuse Act in place, this definition is clear, and the evasive computer hackers are now finding themselves in jail.

### 3.2.1 Patriot Act

After September 11, 2001, the U.S. Congress passed the Patriot Act. This act amended Section 1030 of the Computer Fraud and Abuse Act and provided the government with expanded cooperation of service providers. They now were required to perform tracking of hacking, denial of service, and other computer crimes throughout their networks. When a computer crime is deemed a federal offense, the service provider is required to provide the records indicating where the offending party had initially connected to the service provider's network and what malicious actions they performed.

By now, one must be asking what constitutes a federal computer crime, especially one that is subject to the Computer Fraud and Abuse Act. To constitute a federal investigation, there must be damage or loss that exceeds the average loss within that geographic area. This means that in a large, economically viable city such as New York, this could be as high as $100,000. In a small city out in a rural area, this amount could be as small as $2000 or $3000. An attack that only scans a network and does not compromise any data or create any denial of service would not qualify for a federal investigation.

## 3.3 State Computer Crime Issues

During the late 1990s and into the turn of the 21st century, state governments had increasing problems dealing with small computer crime. This stemmed from a number of incidences where the computer crime was under the dollar limit required to constitute a federal investigation. On top of this comes another issue of pursuing suspects outside the state in which the victim lives. Very rarely is an item purchased online from a Web site owned and operated in the same state as the customer. This means that if someone were to fall victim to an online fraud incident, the state law enforcement agency would have no authority to apprehend a hacker or criminal outside the state in which that victim lived. This has created a massive problem in working together. Some might think that the easy answer would be to have them work together. This may seem true, although once one gets into the details of this, one starts to see the massive amount of small computer crimes compared to the small amount of law enforcement personnel dedicated to them. This has hampered the ability of state agencies to work together. This is because one state might pick a few cases to investigate and choose not investigate others. One state might not investigate a case that another state did investigate. This often happens and leads to the victim and the criminal being in different states. Once that is determined, the investigating state needs the help of the state that did not choose to investigate the case initially. As one can see, this is where the communication problem begins.

# 3.4 Chapter 3 Review Questions

1.  What piece of legislation defines communication as "the transfers of signs, signals, writing, images, sound, data, or intelligence of any nature transmitted in whole or in part by wire, radio, electromagnetic, photo-electronic, or photo-optical systems that affects interstate or foreign commerce"?
    a.  Patriot Act
    b.  Electronic Communications Privacy Act
    c.  Computer Fraud and Abuse Act
    d.  Computer Crime and Abuse Act

2.  What act detailed the use, disclosure, interception, and privacy of electronic commutations?
    a.  Patriot Act
    b.  Electronic Communications Privacy Act
    c.  Computer Fraud and Abuse Act
    d.  Computer Crime and Abuse Act

3.  How much loss must a company have before it can get help from the FBI?
    a.  $5000
    b.  $50,000
    c.  $500,000
    d.  It is based on location and the average loss for that location.

4.  What did the Patriot Act amend relating to computer crime? Choose the best answer.
    a.  Nothing; it was its own act.
    b.  Electronic Communications Privacy Act
    c.  Computer Fraud and Abuse Act
    d.  Computer Crime and Abuse Act

5.  When was the Electronic Communications Privacy Act created?
    a.  1999
    b.  1988
    c.  1985
    d.  1986

6. When was the Computer Fraud and Abuse Act created?
   a. 1999
   b. 2001
   c. 1989
   d. 1996

7. What act was created to help protect U.S. Government computers?
   a. Patriot Act
   b. Electronic Communications Privacy Act
   c. Computer Fraud and Abuse Act
   d. Computer Crime and Abuse Act

8. What act was created to protect U.S. networks?
   a. Patriot Act
   b. Electronic Communications Privacy Act
   c. Computer Fraud and Abuse Act
   d. Computer Crime and Abuse Act

9. Under the Patriot Act, who is required to log and track hacking attempts?
   a. The Department of Homeland Security
   b. The FBI
   c. The NSA
   d. The ISP

10. What regulation, act, or law was put into place due to the lack of clear terms regarding what constitutes a computer crime?
    a. Patriot Act
    b. Electronic Communications Privacy Act
    c. Computer Fraud and Abuse Act
    d. Computer Crime and Abuse Act

11. The FCC has a law that makes running a sniffer and receiving other people's network traffic a crime.
    a. True
    b. False

# Chapter 4

# Wireless Physical Layer Technologies

This chapter looks at wireless physical layer (Figure 4.1) technologies, which encompass the methods of radiating the wireless signal across the air. Many different techniques have been used over the years to send radio waves. When the IEEE got together to create each of the 802.11 standards, it decided on the method it felt would best provide the needed ability to send these radio waves as far and clean as possible.

All the different methods used by the IEEE to send wireless signals today over the airwaves are based on a spread spectrum concept. Wireless networking utilizes this spread spectrum technology at its physical layer to communicate over the airwaves. Spread spectrum technology works by using different modulation techniques over a wider band of frequency than necessary. This makes spread spectrum signals look like noise, which makes them difficult to detect, intercept, or jam. The military has often favored spread spectrum technology compared to narrowband, which sends a signal on a very small frequency band at a very high power. Looking at Figure 4.2, one can see the difference between spread spectrum and narrowband.

Military personnel used spread spectrum technology because they were able to transmit signals over such a wide frequency band that eavesdroppers could not distinguish it from radiant noise. Because of this, the military kept spread spectrum technology a secret until the 1980s when the FCC implemented some rules making it available to the public. The public release of the technology was to encourage research and investigation, leading to more advanced methods of transmission and securing

**49**

| Application |
| :---: |
| Presentation |
| Session |
| Transport |
| Network |
| Data Link |
| Physical |

**Figure 4.1   Wireless physical layer.**

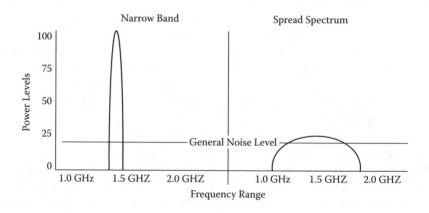

**Figure 4.2   Spread spectrum and narrowband.**

wireless communications. This ended up leading to the 802.11 family of standards, which incorporates spread spectrum technologies.

The FCC has always tried to conserve as much of the frequency bandwidth as possible, only allowing what is needed for radio communications. Using narrowband transmission is how the FCC can conserve frequency and allow for numerous transmissions within a small radio frequency band. Narrowband communication requires a high power output to pass the signal beyond the inherent noise within any given air space. This inherent noise is often called the noise floor. As one might guess, this is particularly prone to interference and jamming because of the small amount of frequency being used. Using a wider band makes interference and the threat of jamming a reduced risk. This is why spread spectrum technology has been adopted by the military as a more reliable and robust means of communication.

The history of spread spectrum technology is a very interesting one. It starts out with Hedy Lamarr, formerly Hedwig Eva Maria Kiesler, who was born in Vienna on November 9, 1913, to Jewish parents. In the late 1940s, she avoided the advancement of the German army by marrying an Austrian arms dealer. This man was one of the leading arms manufacturers in Europe. His name was Mandl and he operated a factory that supplied Hitler in his European campaigns. Hedy took on the role of hostess and entertainer for his guests. In doing so, she often overheard conversations from multiple arms dealers about the inherent problems with jamming torpedoes. She then thought that instead of using narrowband communication, which often guided those torpedoes at that time, they could, instead, use a technique that took the signal, spread it across a large frequency, and switched from one particular frequency to another inside the frequency range.

Hedy, the woman who had learned about the latest in German and Austrian military technology at her husband's plants, met composer George Antheil at a dinner party in 1940. This was where she shared what she knew about the design of remote-controlled torpedoes. Hedy believed the solution was to broadcast the weapons signal on rapidly shifting frequencies. She and Antheil developed a frequency-hopping system by which both the transmitting and receiving stations of a remote-controlled torpedo changed at intervals based on time. They received U.S. Patent 2,292,387 in August 1942, and the U.S. Navy put their research to limited use during World War II.

What makes this even more interesting is who Hedy Lamarr became. She later became a movie star, appearing in many motion pictures. One scene that most defines her career is a ten-minute nude swimming scene in the movie "Ecstasy." This movie made Hedy known as the first women to make a nude appearance on film. Hedy Lamar died on January 30, 2000, in her home in Orlando, Florida, at the age of 86. She had only just begun to benefit from her patent a mere three years earlier.

## 4.1 ISM Spectrum

In 1985, the FCC modified Part 15 of the radio spectrum regulation. This modification authorized the use of wireless network products operating

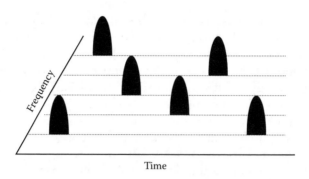

**Figure 4.3    Frequency hopping spread spectrum.**

in the Industrial, Scientific, and Medical (ISM) bands. This free radio spectrum allowed the public, along with the commercial sector, to utilize and produce products that communicate in this spectrum. This opened up a large market for companies to sell wireless products transmitting on this spectrum. If one had a cordless phone in the late 1980s, one can see how this modification affected commerce. Operating on an FCC spectrum usually requires a license. In the ISM spectrum, this holds true to some degree. The FCC requires that it verify products to make sure they do not exceed the maximum power settings. Lucky for us, most products on the market do not operate with enough power to require a license. Any ISM spectrum product must meet certain requirements, such as operation under one watt for transmitter output power.

The ISM spectrum is open for use by anyone as long as the devices operate within the guidelines set by the government. Many devices use the ISM spectrum for other than wireless networks. Microwaves and cordless phones are just a few devices that also share the 2.4 open ISM band. This means they can cause RF interference to any 2.4-GHz wireless networks in the area.

## 4.2 Frequency Hopping Spread Spectrum (FHSS)

Frequency hopping spread spectrum (Figure 4.3; FH or FHSS) is a method for transmitting data on the physical layer of the 802.11 standard. This works by dividing the 83.5-MHz wide channels into a number of smaller frequency slots. Each country has a regulatory domain that specifies the allowed frequency slots. In the United States, the FCC has allotted for 79 channels between the 2.402- and 2.479-GHz band.

When frequency-hopping communication takes place, the signal must transmit on a particular frequency slot for a given amount of time. The time before the signal is required to hop is identified as the dwell time. Once this dwell time has expired, the frequency-hopping network is required to shift the communication into a different frequency slot. Once the dwell time is determined, there is a time delay in the electric circuitry to change the transmission frequencies from one slot to another. During this time, the radio is unable to transmit, creating a dead time (also known as the hop time).

Frequency hopping proved resilient against narrowband interference. When a narrowband signal occupies a particular frequency band within the range of the frequency-hopping system, it will only interfere with the system for short time while it transmits on that particular frequency. What this means is that, during the course of communication, a narrowband signal might be transmitting over five of the 79 channels used, meaning interference points take place when the hop sequence inserts the data on one of the five channels being utilized for the other narrowband signal. With the available 79 channels and the hop sequence only landing on those five channels rarely, one can see the interference avoidance advantage of frequency-hopping systems.

Frequency hopping applied to wireless networking was only able to yield 1- to 2-MB (megabyte) data rates. Frequency hopping was the dominant technology in the 802.11 standards. This was due to the placement of multiple frequency-hopping access points in a given area without the concern of them contending for the available radio frequencies. How wireless networking clients attached to these frequency-hopping networks was through the wireless beacons sent out by the access point. Chapter 5 discusses the beacon further. As applied to frequency hopping, these beacons contain the hop sequences. These hop sequences are what give the cards the ability to shift through the frequency slots and move to the next frequency slot that the access point was going to talk to. This is all based on the hop sequence that is predefined within the IEEE 802.11 standard.

Table 4.1 shows the other allotted channels and frequency bands in countries outside the United States. When transmission takes place, the communication is divided among these frequency slots and a pseudorandom sequence is used to trigger the data to move into a different frequency slot within this 83-MHz band. These timing functions are called hop sequences, which are also regulated.

## 4.3 Direct Sequence Spread Spectrum (DSSS)

The IEEE 802.11 as a physical layer transport method defines direct sequence spread spectrum (DSSS). With the proliferation of 802.11b, some changes were made to DSSS to match the 5.5- and 11-MB data rates

**Table 4.1    Frequency Hopping World Channel Allocation**

| Country | Channel | Range (GHz) | Hop Size |
|---|---|---|---|
| United States | 2 to 79 | 2.402–2.479 | 26 |
| Canada | 2 to 79 | 2.402–2.479 | 26 |
| Britain | 2 to 79 | 2.402–2.479 | 26 |
| France | 48 to 82 | 2.448–2.482 | 27 |
| Spain | 47 to 73 | 2.473–2.495 | 35 |
| Japan | 73 to 95 | 2.473–2.495 | 23 |

written into the new standard. This is sometimes referred to as high-rate wireless. The 802.11 standard has both FHSS and DSSS listed as a choice for physical layer transport. The IEEE had foreseen that FHSS would be a much more difficult protocol to adapt to a higher bandwidth scenario. With DSSS being an easier protocol to adapt to a higher bandwidth scenario, it was chosen.

The reasoning behind originally putting FHSS in the 802.11 standard was for bandwidth and battery life. On an FH network, running multiple access points in the same location to provide bandwidth over the 2-MB standard was a big plus. Also noted by the IEEE was the increased battery life on FH devices compared to DS devices. These reasons are what keep FH in the 802.11 standard, unlike the 802.11b standard, which only chose a single spread spectrum. When the IEEE made 802.11b, it needed to also make it backwards capable or the existing market would not be so forthcoming to adopt this new technology. With higher data rates and backwards capability, 802.11b became a big hit and with it so did the underlying physical transport DSSS. This has made direct sequence (DS) one of the most widely used wireless LAN spread spectrum techniques today.

Direct sequence uses two of the modulation techniques we looked at in chapter one, BPSK and QPSK. Direct sequence is a method of transmitting data across the air on a 22-MHz-wide frequency range. This is done to prevent narrowband noise or interference. This is performed by sending the signal across a large bandwidth of frequency and eliminating any small narrowband noise. When this narrowband interference appears, which is often in short small spikes, it is easily distinguished from a slow-rising or sloping wave like the ones in DSSS. To have these larger, less interference-prone waves, a large frequency bandwidth is needed. This large bandwidth eats up the small amount of frequency allocated in the ISM spectrum.

Governmental bodies regulate allowed channels that can exist within this 2.4-GHz range. Looking at direct sequence channels, one notes that there are 14 available channels. Table 4.2 illustrates that each of these

**Table 4.2   Direct Sequence Spread Spectrum World Channel Allocation**

| Frequency | Channel | Americas | Middle East | Asia | Europe | Japan | Israel |
|-----------|---------|----------|-------------|------|--------|-------|--------|
| 2.412 | 1 | X | X | X | X | X | |
| 2.417 | 2 | X | X | X | X | X | |
| 2.422 | 3 | X | X | X | X | X | X |
| 2.427 | 4 | X | X | X | X | X | X |
| 2.432 | 5 | X | X | X | X | X | X |
| 2.437 | 6 | X | X | X | X | X | X |
| 2.442 | 7 | X | X | X | X | X | X |
| 2.447 | 8 | X | X | X | X | X | X |
| 2.452 | 9 | X | X | X | X | X | X |
| 2.457 | 10 | X | X | X | X | X | |
| 2.462 | 11 | X | X | X | X | X | |
| 2.467 | 12 | | X | X | X | X | |
| 2.472 | 13 | | X | X | X | X | |
| 2.477 | 14 | | | | | | |

channels starts 5 MHz from each other and each has a bandwidth of 22 MHz. This makes these channels often interfere with each other. The proper designing of wireless networks must take into account this inherent overlap and assign proper channel selection. Looking at the channels available in the United States, one notes that there are three nonoverlapping channels: (1) channel 1, which operates at 2.401 to 2.423 GHz; (2) channel 6, which operates at 2.426 to 2.448 GHz, and (3) channel 11, which operates between 2.451 and 2.473 GHz. Placing all these channels on a linear path gives one a better feel for their relation and inherent interference between the selectable channels. A full list of all the channels and their frequencies for North America is listed in Table 4.2.

Looking at Table 4.2, channels 1, 6, and 11 do not overlap each other. The exclusive uses of these channels are recommended in a best practice design. Understanding what causes overlapping channels is a big part of being a successful RF engineer. To understand channel overlap, take a look at the following example. Case 1: you have stuck with the design guidelines and have implemented on channels 1, 6, and 11. Now that you are done, there are some parts of the facility where you have channel 6 overlapping with itself. This creates a large, layer two domain, such as daisy chaining two hubs together. This would make both areas of coverage

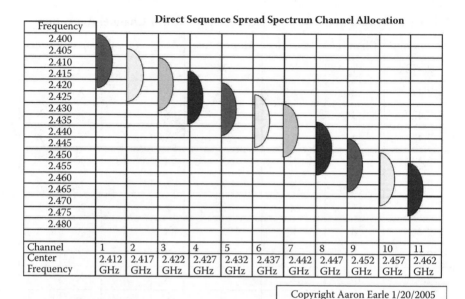

Figure 4.4 Direct sequence spread spectrum channel overlap detail.

the same broadcast domain. This is difficult to avoid and appears multiple times. To overcome this, one can try moving to a four-channel design. This may increase the network throughput, although it is a trade-off between large broadcast domains and slight channel interference. Performing throughput testing using both methods would allow an engineer to adapt the best method to deal with the limited nonoverlapping space. To recap, if one is in a situation where one might need to use a four-channel design, perform some throughput testing to see if channel interface or a large broadcast domain produces better throughput in the particular situation.

Various countries limit the use of the available channels. Looking at Table 4.2, one can see each of the countries and their allowed wireless DSSS channels. In the United States, only the use of channels 1 through 11 are allowed. The United Kingdom can use channels 1 through 13. Finally, Japan uses all 14 channels. Be aware this can complicate matters when designing international wireless local area networks (LANs). This can also be seen when you deal with end devices that travel between these countries. When this happens, one can take two approaches: (1) use 802.11d to dynamically adjust client devices when connecting to an access point in different countries or (2) just design setting the common channels across all access points globally. This can be done by never using channels 12 through 14 in Europe or Japan.

# 4.4 Orthogonal Frequency Division Multiplexing (OFDM)

In July 1998, the IEEE selected Orthogonal Frequency Division Multiplexing (OFDM) as the basis for the 802.11a standard in the United States. OFDM is a wireless technology patented in 1970. As seen in Chapter 1, OFDM works with four modulations: BPSK, QPSK, 16-QAM, and 64-QAM. OFDM works using the allocated frequency band and dividing it into a number of low-frequency sub-carriers. In 802.11a, the 20-MHz channel is divided into 52 such sub-carriers that are 312.5 kHz wide running across eight standard channels. These sub-carriers are orthogonal, meaning every sub-carrier can be separated out at the receiver without interference from other sub-carriers. This is possible because the mathematical property of orthogonal waveforms ensures that the product of any two sub-carriers is zero. This also allows the receiving end to reconstruct damaged portions of the signal. OFDM technology can be seen in broadcast systems such as Asymmetric Digital Subscriber Line (ADSL), Digital Audio Broadcasting (DAB), Digital Video Broadcasting Terrestrial (DVB-T), 802.11a, and 802.11g.

# 4.5 Chapter 4 Review Questions

1. An 802.11a radio uses what technique to transmit its signal?
   a. FHSS
   b. DSSS
   c. OFDM
   d. OFSS

2. The 802.11b standard does not use DSSS.
   a. True
   b. False

3. Which standard listed below supports Frequency Hopping Spread Spectrum?
   a. 802.11b
   b. 802.11g
   c. 802.11a
   d. 802.11

4. Within North America, what number of channels are considered nonoverlapping?
   a. 1
   b. 3
   c. 14
   d. 11

5. The 802.11g standard supports the use of orthogonal frequency division multiplexing (OFDM) in which radio frequency band?
   a. 2.4 MHz
   b. 5.4 MHz
   c. 5.8 MHz
   d. 5.0 MHz

6. Dwell time is a function of which of the following physical layer techniques?
   a. DSSS
   b. FHSS
   c. OFDM
   d. PHY

7. DSSS is a type of what?
   a. Spread surface
   b. Spread spectrum
   c. Modulation type
   d. Network identifier

8. How many channels of DSSS are in use in the United States?
   a. 1
   b. 14
   c. 12
   d. 11

9. How many channels are used in an ESTI-governed location?
   a. 1
   b. 13
   c. 11
   d. 14

10. The 2.4 frequency that 802.11b operates in is called _____.
    a. DSSS
    b. OFMD
    c. UNII
    s. ISM

11. The 802.11a standard is associated with which two terms?
    a. ISM
    b. UNII
    c. 2.4 GHz
    d. 5.15–5.30 GHz

12. How much of the 2.4-GHz ISM band is used when a single access point 802.11b wireless network is enabled?
    a. 11 MHz
    b. 22 MHz
    c. 1 MHz
    d. 20 MHz

13. What is the total amount of wireless bandwidth inside the 2.4-GHz ISM band?
    a. 70 MHz
    b. 11 Mbps
    c. 54 Meg
    d. 83.5 MHz

14. What is the amount of time a channel takes to switch from one channel to another channel on an FH system called?
    a. Hop time
    b. Dwell sequence
    c. Hopping time
    d. Dwell time

15. You are designing an RF network and have channel 6 used. What other channel would you select to use next to channel 6?
    a. 4
    b. 3
    c. 1
    d. 9

# Chapter 5

# Wireless Management Frames

This chapter focuses on the data-link layer (Figure 5.1) of the Open System Interconnect (OSI) model. This is where wireless frames are defined. The data-link layer is the second layer of the OSI model; it takes information and divides it up into frames. These frames are forwarded to the first part of the OSI model, the physical layer, to be transmitted out of the wireless network's antenna. While Chapter 4 focused on wirelesses physical layer technologies and how they work, this chapter explores wireless frames.

When most people think of wireless Ethernet, they think it is very similar to wired Ethernet. This is not the case; Ethernet was created upon 802 just like wireless, although Ethernet is defined as 802.3 and wireless is defined as 802.11. The difference here is how the MAC layer information is handled. In Ethernet, the maximum frame size is 1518 bytes; wireless has a maximum frame size of 2346 bytes. There are some major differences between wireless Ethernet and normal wired Ethernet. This chapter sheds some light on this as it discusses the mechanisms used for wireless frame management. Some of these mechanisms are very similar to wired Ethernet and some are very different.

There are three main types of frames used inside the 802.11b/a/g wireless standards. These frames make up all wireless communications. They are control frames, management frames, and data frames. Control frames control the data flow; management frames allow users on and off the network; and most importantly, data frames move the precious data.

Looking at Figure 5.2, one can see a general 802.11/b/g/a frame and that many parts make up just a single frame. Looking more closely at the

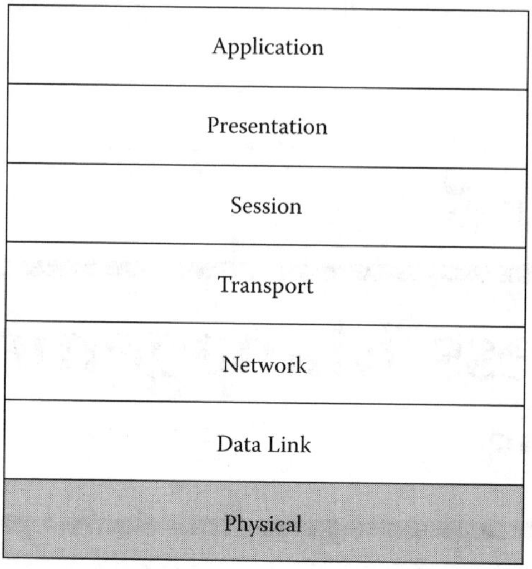

**Figure 5.1   Wireless data-link layer.**

frame's control section, one sees that this 2-byte portion contains a large amount of identification information. Looking at each part, one can see exactly what its function is.

- *Protocol Version.* This is a field to identify the frame as a wireless frame. Currently, there is only one setting for this field.
- *Type.* This part of the frames identifies the frame as a control frame, a management frame, or a data frame.
- *Subtype.* This portion of the frame further identifies itself as one of the various types of control, management, or data frames.
- *To DS.* This field indicates if the frame is destined for a Distribution System, which identifies a wireless network.
- *From DS.* This field indicates if the frame is sourced from a Distribution System.
- *More Fragments.* This field is used to indicate that there are more frames arriving. This means other fragments or portions of this frame will be following this frame.
- *Retry.* This field informs the receiver that the frame has been already sent and that it was informed to retry the transmission.
- *Power Management.* This field is used to identify if a device is in the power save mode. Access points do not go into power save mode and always send or set this value to zero. Clients, on the other hand, do go into power save mode and use this frame indicator.

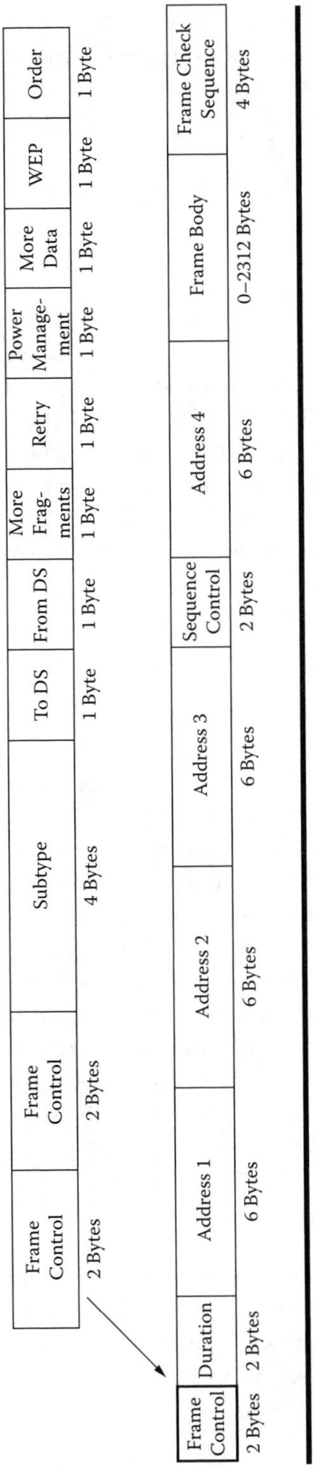

| Frame Control | Subtype | To DS | From DS | More Frag- ments | Retry | Power Manage- ment | More Data | WEP | Order |
|---|---|---|---|---|---|---|---|---|---|
| 2 Bytes | 4 Bytes | 1 Byte | 1 Byte | 1 Byte | 1 Byte | 1 Byte | 1 Byte | 1 Byte | 1 Byte |

| Frame Control | Duration | Address 1 | Address 2 | Address 3 | Sequence Control | Address 4 | Frame Body | Frame Check Sequence |
|---|---|---|---|---|---|---|---|---|
| 2 Bytes | 2 Bytes | 6 Bytes | 6 Bytes | 6 Bytes | 2 Bytes | 6 Bytes | 0–2312 Bytes | 4 Bytes |

**Figure 5.2 Wireless frame detail.**

- *More Data.* This field is used to inform clients that the access point has received multiple data frames for the receiver and has started to fill its buffer with them.
- *WEP.* This field is used to indicate if WEP encryption is or is not being used.
- *Order.* This is a portion of the frame to help with Quality of Service (QoS).

## 5.1 Beacon

Access points identifying themselves and their configuration settings often send out beacon frames. Clients can send out beacons only when in ad hoc mode. These beacons contain the access points' supported data rates, the channel that it is talking on, and if it supports WEP or not. Depending on the settings, it might also broadcast the SSID. The beacon frames are used for wireless clients to understand who can provide them service in a given area. A beacon frame is like an access point screaming out how it is configured so that clients can hear this.

## 5.2 Probe Request

Probe requests are similar to that of a beacon in that they are both used for identification and configuration information. Unlike the beacon, which transmits the identification of itself out over the air waves to any device in that area, the probe is used by the wireless client to locate an access point with the similar identification settings. If the client's probe has the correct SSID, then the two devices determine if their settings allow them to start the authentication process. This probe request contains two main parts: (1) the SSID and (2) the supported data rates. Probe requests are often used to find access points that are not sending beacons. Access points can suppress their beaconing, so a mechanism to find them must also exist. This is where the probe request fits in.

## 5.3 Probe Response

An access point that hears a probe request will respond. It will only respond when the SSID from the request matches what the access point is configured to use. This probe response packet will identify the configuration settings of the access point. The probe response packet will have the following settings: data rates, SSID, and channel information. This is similar to access point beacon, although it will only be sent if a client places a correct probe request that matches identification information with the access point.

## 5.4 Authentication

This section identifies the means of authenticating to a wireless network. How the authentication is accomplished depends on the security settings and the level of enabled security. Authentication must take place before attempting to connect to a wireless network. This frame has a number of identification mechanisms to identify and authorize a wireless device. Looking at the frame in detail, an authentication frame has four main parts, which are unique to it:

1. *Authentication algorithm:* a portion of the frame identifying what security settings it will be using.
2. *Authentication transaction number:* identifies which frames are parts of which authentication transaction. This is used when an access point is serving multiple authentication attempts simultaneously.
3. *The status code:* used to identify why a failure or success occurred.
4. *Challenge text:* a means of security by which a cleartext packet is sent for security purposes.

## 5.5 Association Request

Once a wireless device has successfully authenticated, it can then start the next step of joining the network, called association. To associate, the wireless client must again send its configuration information to the access point. The access point looks at this information and makes a decision of whether or not to join the device to the network. This decision is not based on authentication because that has already taken place before this step; it only verifies that the settings between the two are agreed upon by both parties. These settings are SSID, channel, and power.

## 5.6 Association Response

This frame, which is generated by the access point, identifies to the wireless device that it has been connected to the network. Association response frames are closely related to association request frames. The only difference between the two is a small identification attribute in the frame. Once the client gets this frame, it knows that they have successfully connected to the network and are ready to start utilizing network resources.

## 5.7 Disassociation and De-Authentication

These two frames are the same, with the only difference identifiable through the reason code. This code is embedded in the body of the frame and is used to identify why a client was removed from a wireless network. The reason code is a small 2-byte portion of the frame. This reason code is capable of identifying more than 50,000 reasons; luckily, there are only ten currently specified reasons for this field. The reasons used inside this code are laid out from least significant to most significant, running from 0 bytes as the least significant to 65,535 bytes as the most significant.

Both of these frames are used to remove a wireless device from the network. The main differences between these two frames are the reason for why they removed a device and how or what piece of equipment made the determination to remove the wireless device. The Disassociation message is used by the access points to remove clients that have not talked for a particular amount of time. This allows the access point to clean up after it- self and remove old clients that have disconnected without notice. The De-Authentication frame is sent by an administrator is remove a wireless device from the network.

## 5.8 CSMA/CA

Carrier Sense Multiple Access with Collision Avoidance (CSMA/CA) is the technology that operates at the data-link layer. This second layer of the OSI model responsible for moving frames across its known view of the world or its network segment. This is similar to Ethernet's Carrier Sense Multiple Access with Collision Detection CSMA/CD, which is a second OSI layer protocol to move frames in an Ethernet environment. The difference between the two is more than what media the frames are moving across. With wireless networking having bandwidth at such a premium, a scheme with an acceptance for bandwidth loss was not exactly the best approach. On the wire, losing a small amount of bandwidth to speed up the movement of network frames was considered acceptable. This was where the last two words in the CSMA/CD acronym stand for "collision detection." In CSMA/CD, it detects that a collision took place and resends the frame. On a wireless network losing bandwidth along with time to detect that a collision took place was not acceptable, so another mechanism (called collision avoidance) was utilized. The process of how CSMA/CA works is detailed below, including each of the four critical steps involved in a transmission and how they work together to move frames across a wireless network.

## 5.8.1 RTS

The first step in this process is the initial request. This is called the Request to Send or, for short, RTS. The client computer initiates the RTS when it needs to send data to the access point or through the access point to other network resources. As part of the initial request, the client determines the amount of time he or she needs to complete the transmission. This is set with the NAV bit. The access point will use this to make a decision on how long the client can talk before it has to ask permission again.

Looking at Figure 5.3, one can see that this 20-byte frame has five different parts. First is the frame control section, which is a field identifying the frame as RTS. Next is the duration portion, which signals to the access point the amount of time that the client wishes to have the airwaves for to send information. Next is a section called the receiver address; this address is the MAC of the sending station. Next is the transmitting address section, which is also composed of a MAC address. The final portion consists of the frame check sequence. This is the wireless version of an Ethernet checksum. With the wireless airwaves being less reliable media than standard Ethernet, a different technique was used to send a checksum. This technique works by sending a checksum for the previous frame.

| Frame Control | Duration | Receiver Address | Transmitter Address | Frame Check Sequence |
|---|---|---|---|---|
| 2 Bytes | 2 Bytes | 6 Bytes | 6 Bytes | 4 Bytes |

**Figure 5.3   Request to send frame detail.**

| Frame Control | Duration | Receiver Address | Frame Check Sequence |
|---|---|---|---|
| 2 Bytes | 2 Bytes | 6 Bytes | 4 Bytes |

**Figure 5.4   Clear to send frame detail.**

## *5.8.2 CTS*

Next, one sees the request as it is returned by the access point with a clear to send frame (CTS). This frame indicates how the long the client can talk for before having to close the CSMA/CA process again. This CTS frame consists of segments to the RTS frame. Looking at Figure 5.4, one can see the frame control segment, which announces its existence as a CTS frame. Next is a duration segment, which makes up the amount of time the wireless device has to transmit data. Next is the receiver's MAC address, which is copied from the previous frame's transmitting address, and finally the frame check sequence.

## *5.8.3 DATA*

This section does not deal with any management per se, however, the data section is needed to fully understand what sequence of events takes place when a wireless device wishes to transfer data. Looking at how this plays a role in the entire data transmitting sequence, one can see that the data frame is sent and subsequent data frames continue to be sent until the time allotted by the CTS frame has been used. Once this happens, the data is finished transmitting until a new CTS frame is received with a new period for which the data frame process can resume. To get this new CTS frame, the entire process must be restarted with an RTS. As one can see from Figure 5.5, this frame looks a little different because it is actually a data frame and not a management frame like the RTS and CTS frames.

## *5.8.4 ACK*

Acknowledgment frames are sent by many different networking protocols to identify to the receiver that the data sent was received and was not corrupted en route. This acknowledgment frame is made up of similar sections such as the Request to Send and Clear to Send frames. Looking at Figure 5.6, one can see that, like the other frames, the acknowledgment has a frame control section that identifies it as an acknowledgment frame. It has a duration portion that is used to inform the access point if this acknowledgment is the last one in the current data transfer process. Finally, there is the receiver address section, which is copied from the transmitter's MAC.

## 5.9 Fragmentation

Fragmentation is common on wireless networks due to inherent interference. Its function is not what many people think. Fragmentation occurs

| Frame Control | Duration | Destination Address | BSSID | Source Address | Sequence Control | Payload | Frame Check Sequence |
|---|---|---|---|---|---|---|---|
| 2 Bytes | 2 Bytes | 6 Bytes | 6 Bytes | 6 Bytes | 2 Bytes | 0–2312 Bytes | 4 Bytes |

**Figure 5.5   DATA frame detail.**

| Frame control | Duration | Receiver address | Frame check sequence |
|---|---|---|---|
| 2 Bytes | 2 Bytes | 6 Bytes | 4 Bytes |

**Figure 5.6 Acknowledgment frame detail.**

to enable the wireless media to be able to send more data. This is done with the use of smaller transmission. With wireless being such an unstable medium, sending large frames and then, in the event of corruption, having to resend them takes more time than sending and resending smaller frames. Because of this, fragmentation is often performed to speed up the transfer of frames. It is purposely applied to break up the transmission into small pieces that can be resent very quickly. This is important because the wireless medium is only allocated to a single user for a set amount of time and then everyone must contend to use it.

This fragmentation can cause problems as well; fragmentation creates overhead and the processing time requires splitting the frames and reassembling them on the other end. Setting fragmentation can be a tricky option. One needs to make sure it is set between having it over-fragmented and causing processing issues, or having it under-fragmented and causing retransmit errors. Usually, access point manufacturers have tweaked this number to its most optimal setting, although in some cases it may need on-site adjustment.

## 5.10 Distributed Coordination Function

The distributed coordination function (DCF) is used to poll the media or, in this case, the wireless to see if someone else is using it. As seen above, before one can send any frames, one needs to make sure no one is using the network. The DCF performs this by examining the entire MAC header of frames that are sent. When another machine wants to use the network, it requests access, stating how long it needs the network for its own use. When it does this, its frames are tagged with this indication, called the network allocation value (NAV).

## 5.11 Point Coordination Function

The point coordination function (PCF) has the same goal as that of the DCF, although it handles it differently. In the PCF, a device tells the access point that it can respond to polls. This means that PCF can only work within a wireless network controlled by an access point. Once the access point is aware that the device can support polls, it will poll the entire wireless portion, asking if clients need to send any frames. This provides the device that needs to transfer information with the ability to do so. The PCF must function in unison with the DCF. This is not the other way around; the DCF can function without the PCF.

## 5.12 Interframe Spacing

To perform management functions and allow for time synchronization, a preset interval where nothing communicates is required. This is termed "interframe spacing." This value is set into all the devices as part of each 802.11 standard. This value allows for access point management and the prioritization of network traffic.

There are four types of interframe spacing:

1. Short interframe space (SIFS)
2. Point coordination function interframe space (PIFS)
3. Distributed coordination function interframe space (DIFS)
4. Extend interframe spacing (EIFS)

## 5.13 Service Set Identifier (SSID)

When looking at the Service Set Identifier (SSID), one sees that there was never any intention of having it perform any type of security measure. The SSID's main purpose is for network identification, as the name states. When a client end device connects to the network, it must have an identification setting to allow it to know what network to connect to and operate on. When the wireless standards were created, the IEEE had the foresight to realize that there may be more than one wireless network within range. This led to the creation of the SSID as a means to differentiate one wireless network from another. Today, with the massive amount of wireless networks, this has become a necessity. This is what the SSID is for: network identification.

The SSID can also be used to create multiple virtual wireless networks. This is very similar to VLANs, which are used in the wired world. Having multiple wireless networks does mean that everyone is still sharing the same air space, although they are on their own wired subnet. This is often used to accommodate guests and allow for different security levels. This can be seen in a situation where some older devices could not support the advanced security standards. In this situation, two wireless VLANs could be created: (1) one with the advanced security open to go anywhere in the wired network and (2) the other one that has support for the older weaker method of the security. The weaker method of security would also have some other wired security methods such as ACL or IDS, or some other added security mechanism to balance out the higher wireless security risk it brings.

## 5.14 Chapter 5 Review Questions

1. What are the three main wireless frame types?
   a. Control frames
   b. Management frames
   c. Data frames
   d. Session frames

2. When a wireless network is using an access point, which device sends the beacons?
   a. Access point
   b. Wireless laptop
   c. Authentication server
   d. User

3. Which layer 2 protocol does 802.11b use?
   a. 1CSMA/CD
   b. RTS/CTS
   c. DSSS
   d. CSMA/CA

4. What is the difference between a Disassociation and De-authentication frame?
   a. Frame TTL
   b. Frame size
   c. Reason code
   d. Protocol field

5. CSMA/CA is a _____ process.
   a. One-step
   b. Two-step
   c. Three-step
   d. Four-step

6. To perform management functions and allow for time synchronization, a preset interval where nothing communicates is required. What term does this describe?
   a. DCF
   b. PCF
   c. CTS
   d. Interframe spacing

7. Which step takes place first: authentication or association?
   a. Authentication
   b. Association

8. CTS/RTS is part of what?
   a. CSMA/CA
   b. CSMA/CD
   c. WECA
   d. IEEE

9. Which layer 2 protocol does Ethernet use?
   a. CSMA/CD
   b. RTS/CTS
   3. DSSS
   d. CSMA/CA

10. What is the SSID used for?
    a. Security
    b. Identification
    c. Information exchange
    d. Network management

11. The distributed coordination function (DCF) is used to _____ .
    a. Poll the media for network identification
    b. Poll the media for security
    c. Poll the media for network availability
    d. Poll the network for any changes

## Chapter 6

# Wireless Local and Personal Area Networks

What are wireless local area networks (WLANs), and how do they work? This chapter explores what makes up wireless networks, how they are set up, and what kind of hardware is involved. It then looks at how the market has adapted to creating many wireless standards for interoperability between multiple vendors. This chapter introduces the main wireless standards and describes how they work and what differentiates them.

At first, wireless networks were low data rate, proprietary installations that were used only for the most needed mobile applications. This was common in warehousing and other inventory management solutions. Today's wireless networks have gained momentum in a number of vertical markets such as healthcare, education, retail, manufacturing, warehousing, and more. This growth is directly related to the fact that wireless networks give users the ability to remove constraining wires. This freedom from wires allows users to work anywhere there is a network signal present. Wireless networks bring massive gains — not only in productivity, but also from reduced cabling and fast client relocation. All these advantages have made wireless networks pop up all over the corporate landscape and subsequently created this fast-growing industry.

Flexibility is a major reason that wireless networks have become so popular. Just looking at historical buildings gives us an example of this. Once a building is deemed historical, running wires through it can quickly become an unacceptable option. With wireless networks, no wires are necessary; a user just has to plug into an access point and he is set to go. Without having to drill holes for wires, these historical buildings can keep their old-world look and feel.

Another way the flexibility of wireless networks is useful is in areas or buildings not owned by the occupant. In this case, holes cannot be drilled into the walls to install wire runs. Wireless allows one to set up the access point and connects all the needed information systems via a wireless connection.

Disaster recovery is another area where the flexibility of wireless plays a key role. When major damage impedes the ability to hang cables, using wireless can help keep a workforce connected. This can be seen in events such as the tornado that hit the General Motors Oklahoma Assembly Plant in 2003. To help speed up the construction effort, wireless bridges were put in connecting temporary trailers to the GM infrastructure. This gave workers the ability to be connected to the communication tools they needed, thus allowing them to rebuild the plant faster.

Recently, the phone companies have also jumped on the bandwagon, releasing pay-for-use wireless networks located in high-traffic areas. These are called hot spots and can be located in airports, hotels, coffee shops, and many other prime places throughout the world. Just connect and input a credit card or subscription information, then surf the Web, e-mail, or VPN into your corporate network. With market research companies like Gartner reporting hot spot availability going from a little over 1000 in 2001 to 130,000 in 2004, we are likely to see hot spots covering a large portion of heavily populated cities within a few years.

## 6.1 Ad Hoc Mode

Wireless can operate in two main modes. The first one is called Ad Hoc mode, or Independent Basic Service Set (IBSS). This is a peer-to-peer wireless network. This means that it does not have an access point controlling the conversation. This mode is popular for small deployments where the number of users is less than five. In this method, the access point that usually manages the conversions is gone and the clients send beacons to each other instead. From Chapter 5 we learned that this is the only case in which a client would send a beacon frame. These beacons contain a timer synchronization function (TSF) to ensure that the timing is correct. This function is usually handled in the access point. Figure 6.1 shows what an Ad Hoc wireless network looks like. As one can see, there is no access point. This method of wireless was created primarily not to circumvent access points, but rather to allow two or more computers to

**Figure 6.1   Ad Hoc or Independent Basic Service Set (IBSS) diagram.**

**Figure 6.2   Infrastructure Mode or Extended Basic Service Set (EBSS) diagram.**

share each other's resources when together. This helps traveling teams and workers exchange ideas and information more easily in locations where networks do not exist.

## 6.2 Infrastructure Mode

The more commonly adopted mode for wireless is the Infrastructure mode, which is called Extended Basic Service Set (EBSS). This is the main type of wireless network. In an EBSS, an access point controls all traffic. Setting up a wireless network in this category requires a piece of networking equipment referred to as an access point. This access point is where the Ethernet data is converted into a wireless signal that is then transferred out through the access point's antenna. To hear and understand this signal, a wireless network interface card is needed. This card has a small antenna inside it and can hear the wireless signal and transfer it to the computer. This mode is seen in Figure 6.2. This allows a group of people to connect to the network in any area covered by the access point's signal range.

Wireless Bridge

**Figure 6.3   Wireless bridge.**

## 6.3  Bridging

Wireless networks can also extend existing wired networks (see Figure 6.3). This is most often found in bridging applications of wireless. Bridging can be used for a variety of reasons, including connecting campus buildings together, connecting nearby businesses together, or supplying last mile service as an ISP. When wanting to connect two buildings together, instead of paying the high costs of leased lines, one can build a wireless bridge. The ability to build a bridge requires that the buildings be no more than ten miles apart. However, technology and cost can overcome this distance limitation. For example, satellites can be used to bridge any two buildings in the world. Usually, for normal application, ten miles is acceptable. The bridge can also be applied for companies in the same vicinity that do business with each other. These businesses might need to connect together their inventory systems. In the case of the last mile, the ISP will install a wireless bridge on the roof of a home or business and connect it to ISP's local office. This provides Internet services via wireless and saves the ISP from incurring the cost involved in running wires to the home or business.

## 6.4  Repeater

A repeater (Figure 6.4) is an access point that supports a feature allowing it to use wireless instead of wired Ethernet to connect to the backbone. When this is done, the access point that is acting as a repeater has no physical connection to the network. It is connected to the wired network

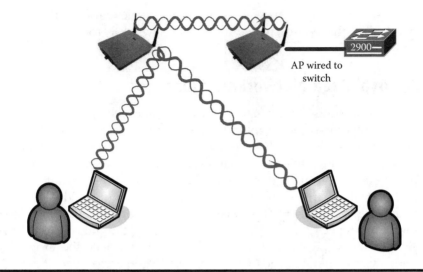

AP wired to
switch

**Figure 6.4   Wireless repeater.**

through a wireless connection to another access point that is wired into the network. This is common for large places that do not have the ability to run cabling and are too large to cover with a single access point. The downside to using a repeater is that all wireless bandwidth for the repeating access points and the one serving the connection to the wire are all the same. This means that everywhere there is a repeater, all clients connected to the repeater, and the other access points are contending for the same amount of limited bandwidth.

## 6.5 Mesh Wireless Networks

Some ISPs, manufacturers, and a number of municipalities have started to use wireless mesh networks. Just like the Internet where routers have many paths to take to get to point A from point B, wireless mesh networks use similar routing protocols to create one highly available network. As seen with hot spots, these networks provide similar functionality on a much larger scale. This form of wireless has been primarily used in city-wide or other very large wireless networks. Its main advantage is that not every access point must be hard-wired into the network. As with repeaters, some access points can function in this manner while others have direct connections into the network. Many cities have started using these types of networks to provide wireless Internet access similar to garbage disposal or other city services covered under taxation. These new types of networks have started to create new access points that make routing decisions like a normal network router as well as providing wireless access. These new network types have also created the need for wireless routing protocols.

## 6.6 Local Area Networking Standards

Once the value and cost reduction of wireless networking became widely known, many companies wanted to start using a wireless network. This created some problems at first because wireless networks were manufacturer specific; and if that manufacturer went out of business, so did the wireless system. To make things worse, all the end devices needed to be changed if the wireless network was upgraded or moved from one manufacturer to another. This was because all wireless networks were manufacturer specific and proprietary. This made the industry push the IEEE to make some wireless standards and help facilitate the growth of wireless with common standards that allowed various manufacturer cards to work with various manufacturer wireless networks.

### 6.6.1 802.11

In 1997, the IEEE standardized the 802.11 standard. This was the first WLAN standard accepted by multiple vendors as a true industry standard. The IEEE defines how different types of media carry data. The scope of the IEEE is broad; it writes standards for Ethernet, fiber optics, Token Ring, wireless of course, and many others. The 802.11 standard defines the Media Access Control (MAC) and physical (PHY) layers for a wireless network. Above these levels, there is the network layer, which has IP as a well-defined standard. Why none of these standards accounts for IP is that IP is non-caring to what media it is passing over.

These networks operate on two physical layers: (1) direct sequence spread spectrum (DSSS) and (2) frequency hopping spread spectrum (FHSS). Each uses a different method of transmitting wireless signals across the airwaves. DSSS uses a wide, single, statically defined channel that is preset in the access point. On FHSS or FH, the access point and the client negotiate a hop sequence, which is used to allow the signal to switch between small slices of frequency in the 2.4-GHz range that wireless 802.11 has defined as usable.

The MAC layer has been standardized to help contend with the interference and excessive loss of frames compared to Ethernet. The 802.11 standard has a maximum data throughput limit of 2 MB. This hindered desktop use of wireless networks. With 100-MB networks on the market and 1-GB networks less than a year away, desktop users were not willing to work with a fraction of their old 10-MB networks, which they fought so hard to have upgraded. This limited 802.11 wireless networks and made mainstream wireless network use a pipedream. These networks were widely deployed for inventory tracking and retail point-of-sale machines. The low data rate was not a problem for small hand-held scanners and other applications. These applications grew because they now were mobile and could better improve worker productivity.

### *6.6.2 802.11a*

In 1999, the IEEE group successfully standardized the 802.11a standard. The 802.11a and 802.11b standards were created at the same time. What delayed the arrival of 802.11a products versus 802.11b was that 802.11b products were already shipping as the IEEE standard was being finalized. Many 802.11b products were released well before 802.11a. This happened because many companies already had performed the research and development involved with 802.11b products and not 802.11a. Once the 802.11b products were complete and being sold, the research and development focused on 802.11a products.

The 802.11a standard allows for 54 MB of data to be used on the airwaves. The physical layer technology Orthogonal Frequency Division Multiplexing (OFDM) is used to transfer the data into radio waves. The next technology, which controls the flow and management of the data, is the Carrier Sense Multiple Access with Collision Avoidance (or CSMA/CA).

The 802.11a standard uses 52 sub-carriers, four of which are used for monitoring, path sifts, and inter-carrier interference. Depending on where in the allowable frequency range one places the 802.11a device or radio, one must be aware of government requirements before transmitting. There are three sets of regulations with respect to power and antenna use with 802.11a. All three are called the Unlicensed National Information Infrastructure (UNII) spectrum and are identified as UNII 1, UNII 2, and UNII 3.

The UNII 1 specification calls for indoor use only and operates in the 5.15- to 5.25-GHz range. It uses four of the available eight 802.11a channels. When using an 802.11a channel in this frequency, one must only provide it a maximum power of 40 milliwatts (mW). One big thing to note about UNII 1 is the requirement for fixed antennas. This means antenna options are not allowed unless they are affixed to the radio by the manufacturer.

Looking at UNII 2, one sees that it has four channels and operates in the 5.25- to 5.35-GHz frequency range. There is a limit on its transmitting power, which is restricted to 200 mW. The UNII 2 spec allows for the use of external antennas. Most 802.11a equipment today uses a combination of UNII 1 and UNII 2; when using these two frequencies together, one must follow the UNII 1 rule about fixed antennas.

New changes have taken place in the FCC regulations allowing more channels and different rules. In February 2004 the FCC made changes that allow for the use of 11 new channels. This will give 802.11a designers 23 available channels. Another primary carrier of the frequency is a type of radar. What this means is there is another carrier or user operating in this frequency range. In this case this primary user is radar; because of this a mechanism must exist to limit the amount of interference. To help keep these two signals inference free, the IEEE created a standard called 802.11h. This standard is explained within this chapter. This standard allows for the ability to sense the usage of radar in a given area and change the access point's signal away from those channels. The FCC mandate that any 802.11a users that want to use the new 11 channels must be 802.11h compliant.

The UNII 3 frequency range jumps a little ways down; unlike the consecutive UNII 1 and UNNI 2, UNII 3 is not consecutive. The UNII 3 range is 5.725 to 5.825 GHz; this is interesting because UNII 3 somewhat overlaps the ISM upper frequency. This does not affect any 802.11 products in the ISM range because they run in the 2.4 range of the ISM. Some of the new 5.8-GHz cordless phones might have similar problems that 802.11 faced, although 5.8 GHz does not interfere with UNII 1 or UNII 2, which are the most prevalent 802.11a wireless transmission methods.

The 802.11a standard has one of the smallest ranges in the 802.11 family. Looking at the maximum range for data rates above the 11-Meg 802.11 standards, one sees that it is only 90 feet. With 802.11a, one has the ability to place multiple access points in the same area, overlapping their coverage cells to provide higher bandwidth. This allows one to have more users per area, sharing a larger amount of bandwidth. This is possible because 802.11a allows for greater channel usage without running into overlapping channel issues that plague DSSS systems.

### 6.6.3 802.11b

The IEEE created the 802.11b standard at the same time (1999) as the 802.11a standard. This standard arose from the need to have greater bandwidth than what was available in the then-current wireless 802.11 standard. This new standard was required not only to have higher bandwidth, but also to be backwards compatible with 802.11. This was done so the many companies with existing 802.11 networks could more easily achieve migration. Moving from an 802.11 network to an 802.11b network was possible because of this compatibility requirement.

The data rate in 802.11b was called out as a maximum of 11 MB, a large difference from the existing 2-MB data rate available in the 802.11b standard. This greater bandwidth has helped many companies take their wireless systems from inventory management applications to desktop computing. This also sparked a new way to market wireless as a low-cost competitor to wired networks. With these incentives, many organizations upgraded their systems to support faster data rates.

The 802.11b standard defines the way wireless networks operate at the physical layer and the logical or data-link layer. In the 802.11b physical layer portion, only a single spread spectrum technique was used. This is unlike in 802.11, in which there is a choice between Direct Sequence Spread Spectrum (DSSS) and Frequency Hopping Spread Spectrum (FHSS). DSSS is what 802.11b uses as a physical layer transport. This means that 802.11b uses a single preset 20-MHz channel operating in the 2.4 ISM frequency range when transmitting. DSSS was chosen over FHSS because it was determined that FHSS would be difficult to expand beyond the 2-MB range.

Now, looking at the MAC layer of 802.11b, one uses the Carrier Sense Multiple Access with Collision Avoidance (CSMA/CA) as a layer two transport. This portion of the 802.11b standard handles the movement of frames on a network segment. Frames control the flow of information and carry data. When creating the 802.11b standard, the IEEE had to define how frame management would take place and subsequently created CSMA/CA. The 802.11b standard is one of the most widely used wireless standards today. Most wireless systems are using this standard. Most all laptops ship today with built-in 802.11b or 802.11g wireless capability.

### 6.6.4 802.11c

This was an older IEEE standard group working on MAC bridges. This working group created the 802.11c standard, which details how wireless bridges operate. Almost all the details of this standard only apply to designers and developers of the access point or bridge itself. Most installers or wireless network designers need not be familiar with the 802.11c standard other than to know it exists and that it details bridge operations.

### 6.6.5 802.11d

The IEEE completed the 802.11d standard in 2001. It addresses the need for access points to have the ability to inform client cards of what regulator domain they are located at and what rules apply for that location. As seen in Chapter 4, many rules exist detailing the operation of wireless networks.

These rules only apply to the country in which the equipment is operating. In an effort to make interoperable products between these rules, a method for identifying what rules apply when became necessary. This is where the 802.11d standard comes into play. This standard gives business travelers the ability to use their wireless network cards in North America one day and in Europe the next day.

The client card manufacturers also encouraged this standard. They wanted to provide a means to produce a single client card that would be able to carry over into other regulator domains. The 802.11d standard helps these manufacturers by not requiring them to create different client card versions for each domain in which they are going to sell them. This also helps normal business travelers because they do not have to carry around multiple client cards.

### 6.6.6 802.11e

In the 802.11e standard, one sees the quality of service. This is critical in time-sensitive communications such as voice or video. Having a wireless network, QoS aware network is required to make quality calls on cordless VoIP phones. Most of the Wi-Fi phones on the market today use a proprietary method of QoS. Currently, the 802.11e standard is in draft. Once the draft becomes finalized, these proprietary mechanisms of QoS will adopt 802.11e.

### 6.6.7 802.11f

This one standard should have been inserted directly into all 802.11 flavors. This has always been an amusing area in the wireless industry. When the 802.11 standard came about, interoperability was finally here. If a product was built to the 802.11 standard, it should work with other 802.11 products. This only held true when one meshed two access points made from the same manufacturer. If one took two products made by different manufacturers and tried to create one network, there would be no interoperability. The reasoning behind this was different roaming algorithms. Each manufacturer had determined a different method of roaming. It was not their fault that the 802.11 standard called for required support for roaming; unfortunately, they never defined how in the standard. To note here; most people who create a wireless network from two manufacturers think that their clients are roaming, when in fact they are going through the entire process of disconnecting and connecting to each manufacturer's access point. The big difference here is time and state. In a roam, the time a client is not connected to the network is very minimal, so minimal that it most likely will not interrupt any applications running. In the event that the client actually disconnects and reconnects, many applications will lose connectivity and produce various issues.

The 802.11f standard provides a standard for roaming. This allows companies to create products that can seamlessly roam from one to another. This standard was championed by Cisco systems and was finalized in 2003. It specifies a protocol called the Inter Access Point Protocol (IAAP), which is a set of defined rules for how an access point should hand off clients and facilitate roaming. The IAAP is a multicast-based protocol that shares wireless connection information with all other access points on the same subnet. The ability of IAAP to move off a subnet is not possible due to a missing TTL field. Thus far, not many other companies have created 802.11f-compliant products, even with the standard fully ratified.

### 6.6.8 802.11g

In 2003, the IEEE approved the 802.11g standard. When 802.11a came about, the IEEE recommended using a physical layer standard called OFDM. This allowed for a single channel that could be spilt it into multiple sub-channels, providing the capability to have greater data bandwidth on the same amount of frequency bandwidth. When 802.11g was conceived, the IEEE chose this OFDM technique for physical layer transport. This physical layer transport was applied to the 2.4 ISM, unlike 802.11a, which applied it to the UNII spectrum. This was done to be backwards compatible with the 802.11b standard. This means that the 802.11g wireless standard is capable of a bandwidth of 54 MB and can support 802.11b clients.

Along with the added bandwidth of 802.11g comes a small loss in cell size compared to 802.11b. The 802.11g standard is very close to 802.11b from a coverage aspect; however, to get the two within ten feet, one needs to adjust the power settings of each radio. Why would one need to get these two standards so close? Most wireless devices today are still using the 802.11b standard. Although most laptops are moving away from the 802.11b standard and coming straight from the manufacturer with built-in 802.11g cards, they are not the most widely used wireless devices. Hand-held scanners, fork truck terminals, and other inventory management devices have wireless laptops, tablets, and PDA units outnumbered. This is because of their wide usage and deployment since the 802.11 days and even before that running on 900-MHz systems.

Looking at B/G (i.e., 802.11b/802.11g) network coexistence, there are a number of limitations when conforming to the standard. All 802.11b wireless access points run in the same frequency as the 802.11g access points. This means the radios themselves are capable of running 802.11b or 802.11g because they are both in the 2.4 spectrum. However, this does not mean an access point can switch between 802.11b and 802.11g without any hardware changes. The radio portion of the access point just propagates the signal to the antenna, regardless of how it is structured. A new or updated chipset is needed to change from DSSS to OFDM. This *must be installed into any access points* one wants to upgrade. Remember that all wireless access point vendors have different methods of upgrading, or, in some cases, replacement. One might need to purchase a simple expansion card or one might have to change out the entire access point. The 802.11g standard has a requirement to be backwards compatible. This means 802.11g must be backwards compatible with 802.11b. This backwards compatibility comes with a cost to network performance.

When looking at B/G coexistence, one finds that when introducing an 802.11b client to an 802.11g infrastructure, it will default to an 802.11b wireless network. This means that if one allows 802.11b clients into an 802.11g network, it will lower the available bandwidth to 802.11b levels. This is because the access point has only one radio supplying two different physical layer techniques to clients. This was created to help companies move from 802.11b to 802.11g. Once all clients are 802.11g, the 802.11b data rates can be turned off and only 802.11g clients will be able to connect to the network. Until then, the 802.11g network most realistically always operates at 802.11b data router.

### 6.6.9 802.11h

The 802.11h standard is looking at using 802.11a and developing the ability to self-tune, and moving away from congested channels. This will allow for easier setup and less administration. It plans to accomplish this by incorporating two techniques: (1) Dynamic Frequency Selection (DFS) and (2) Transmit Power Control (TPC). As one can see from the names, DFS changes the frequency or channel of an access point in the event that there is interference on that channel. The TCP portion of 802.11h controls the power output on each access point. This will allow each access point to know if it needs to increase power or lower power based on what it can hear from other access points in the area. These two techniques each handle changing the wireless channel or power in the event that interference occurs. This is being done today with products whose only purpose is to perform this type action. The IEEE would like to incorporate this functionally into a standard that each access point would utilize. The 802.11h standard is a move that the industry is making to create wireless networks without needing to design.

## 6.6.10 802.11i

The 802.11i standard is a security standard that can apply to other 802.11 standards. It involves many changes, including the addition of advance encryption cipher AES and better key management functionality. After the many problems with the security built into the 802.11/a/b/g standards, a new security method needed to be developed. The IEEE created a group and finalized the 802.11i standard in 2004. For more information and details, refer to the 802.11i section in Chapter 12. For now, understand that the 802.11i is a security-related standard that runs on top of 802.11b/a/g, providing a more secure access method than the security defined inside each of these standards.

## 6.6.11 802.11j

This standard is for use in Japan only. It defines the physical and MAC layer communications for systems running in the 4.9- to 5-GHz range. As of early 2005, the standard is still in draft format, with the working group still finalizing it. One interesting thing about this standard is the fact that North America has utilized this frequency exclusively for public safety networks. This may lead the 802.11j standard to become widely used in North America for police, fire, and other public safety needs. The United States is watching how Japan will use this standard and will evaluate it for public safety use in the United States.

## 6.6.12 802.11n

Looking forward, the IEEE has started working on the 802.11n standard, which will be the replacement for 802.11g when it comes out. One interesting thing is the requirement that has been thrown around to support both 802.11g and 802.11a with the same device. Since the two standards are in two separate frequencies all together, in order to conform to the standard the access point will most likely have dual radios. Most of the Pre-N shipping equipment today is 802.11b/g backwards compatible and not 802.11a. This standard has been identified to operate at a minimum of 100 Mbps. To achieve faster data rates a new technique known as Multiple in Multiple out (MIMO) has been chosen for use. This works by having multiple signals sent and multiple signals received. In Chapter 8, we will talk about antenna diversity displaying and why this helps RF communications. MIMO goes further than antenna diversity by not only providing multiple antennas for receiving but multiple antennas for transmitting.

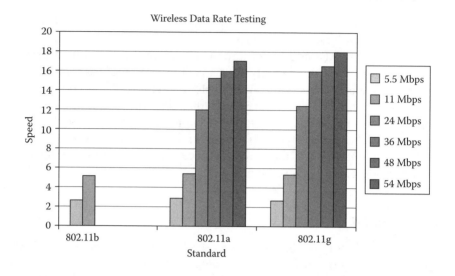

**Figure 6.5  Real wireless data rates.**

## 6.6.13 Real-World Wireless Data Rates

In looking at the many wireless IEEE standards, one sees that each standard outlines what their related data rates are. This number is not possible on normal networking protocols; they are roughly the top speed possible to send simple information through the device. The overhead of intelligent information that keeps track of the transmission takes overhead that the manufacturers and IEEE did not want to use in their top-speed testing. This means that none of the overhead of IP and TCP or UDP is part of this number. Most people will refer to this number as marketecture. Real-world testing reveals a major data rate difference from what is commonly relayed. Figure 6.5 outlines the marketed data rates and the real-world data rates. These data rates have been recorded in a computer laboratory under optimal conditions. This should make the real-world data rates as close to the real maximum data rates possible. It would be difficult to have a better connection in the field than in a controlled environment such as a laboratory. Another interesting fact is the amount of actual data rate gained between the 36-MB and 54-MB data rates; as one can see, this is very small. In some cases, the advantage of the 54-MB cell's data rate outweighs the overhead required to support data rate roaming. What this means is that, in many cases, the 54-MB data rate is actually turned off to prevent excessive, unneeded roaming between the 48-MB and 54-MB data rates.

# 6.7 Personal Area Network (PAN) 802.15

The fight has been on for quite some time now as to how electronics makers can get away from cables. We have seen many cable types used to try to make the interconnection of devices as easy as possible. Take the USB, for example; this cable can be used for so many different things that an entire book could be devoted to it. For some time now, the push has been on for electronics makers to lose the wires. There is no reason why anyone would need to hook his or her radio up to a TV or a digital camera to a computer with wires. Using WPAN (wireless personal area network) technologies allows for the removal of cluttering wires, thus creating wireless interconnections to all support devices.

The market has started to head in this wireless cabling direction for many devices. In doing this, many manufacturers are trying to develop a solution to get rid of the cables. As a direct result of this, there are currently a number of WPAN technologies. This section explores some of the most commonly known standards that make up these technologies.

## 6.7.1 Bluetooth 802.15.1

Bluetooth is a specification for wireless personal area networks (PANs) formalized by the Bluetooth SIG in 1999. It was originally developed by Ericsson, who was a member of SIG with IBM, Intel, Nokia, and Toshiba. The protocol operates in the license-free ISM band at 2.4 GHz, which is globally considered a free, open frequency band. It reaches speeds of 723.1 K a second and has a range of ten meters. It operates on Frequency Hopping and Time Division Duplex hopping 1600 times a second. Devices that are Bluetooth enabled will communicate with each other and find out if they need to be on the same network or not. Once two or more devices realize that they need to be on the same network, a piconet is formed. They know this by a predefined address setting either created by the user or input into the device by the manufacturer. This piconet is used to synchronize each participating device so that they use the same hopping sequence.

Some of Bluetooth's more predominant uses include wireless headsets for mobile phones. These can save time and lives while in use in a car. Another common use of Bluetooth is for wireless keyboards and mice. Recently, new uses for Bluetooth have emerged. One such use is to replace wires in home theater systems. Some way-out uses for Bluetooth that may become commonplace in the near future include some initiatives that Motorola is currently working on. The first one is a partnership with snowboard and clothing manufacturer Burton. Motorola wants to include Bluetooth-enabled speakers and microphones in Burton clothing and helmets. This would allow snowboarders to listen to an MP3 player or use a cellular phone located deep in their pockets with their helmet. Placing a speaker in the helmet and a small number of push-buttons on the sleeves of clothing would allow these Bluetooth-enabled pieces of clothing to place calls and listen to music from any Bluetooth-enabled device. Another partnership that Motorola has started is with sunglasses manufacturer Oakley. The sunglasses will allow the user to listen to MP3 and take cellular phone calls from any Bluetooth-enabled device.

## 6.7.2 Infrared (IR)

Infrared (IR) is not considered a WPAN technology and is not part of the 802.15 family. It is listed here because it has a very small distance limitation, which qualifies it as a personal area network (PAN). Interestingly enough, IR was originally defined in the 802.11 standard. The standard IR has allowed us to connect many devices together quickly and easily for small file transfers. IR is widely deployed on laptops, cell phones, PDAs, and many other hand-held mobile devices. Many people have looked at the security risk of IR; although it has very limited range with almost virtually no security, the risk is low. The risk is so low because of the range factor; one needs to be within one or two meters of another device. This allows anyone to easily see someone who would be trying to attack his or her IR device.

## 6.7.3 Ultrawide Band 802.15.3

The ultrawide band (UWB) is very similar to how it sounds. A short-range communication technology transmits its signal across a wide frequency range. This wide range is also where other signals would be located. UWB is able to prevent itself from affecting or being affected by these other signals by allowing such a wide band. As discussed in Chapter 4, narrowband communications are easily detected and avoided when transmitting a signal at low power over a wide frequency. This is the goal of UWB communications. Currently, there are ongoing talks concerning uses for UWB. Companies have started to develop products used for data transmission, location, and even warfare. The military has been interested in UWB because of its low power and resilience to jamming.

Currently, the UWB standard is being headed up by the IEEE. They have created the IEEE 802.15 WPAN High Rate Alternative PHY Task Group 3a (TG3a) to standardize UWB. This standards body currently has two proposed standards for UWB, and is in the process of removing one of them. Once a choice has been made, that version will become the true

UWB standard. Some of the details of each of these proposed standards are listed below.

For a quick overview of what to expect from UWB, one can count on very high data rates, upwards of 500 Mbps standard. One standard has UWB capped at 1320 Mbps and the other has it capped at 480 Mbps. Both will use the unlicensed 3.1- to 10.6-GHz UWB. How the modulation is handled is where each of the two proposed standards have the most differentiation. One calls for using direct sequence spreading of binary phase shift keying (BPSK) and quaternary bi-orthogonal keying (4BOK) UWB pulses. The other calls for using orthogonal frequency division multiplexing (OFDM) with 122 sub-carriers modulated using quadrature phase shift keying (QPSK). Either way this standard goes, it will be a long-awaited replacement for electronic device cabling.

### 6.7.4 ZIGBEE 802.15.4

Zigbee was created from the 802.15.4 standard. This standard was published in May 2003 to create a low data rate solution that could have long battery life. The battery life is dependent on the device it is servicing. The 802.15.4 standard calls for multi-month to multi-year battery life. Another key function of the standard is its ability to operate with very little complexity.

The 802.15.4 standard, like most IEEE standards, only addressed the physical and MAC layer portions. The Zigbee Alliance came in and addressed a standard for the network layer through the application layer. This included security management, topology management, routing, MAC management, discovery protocol, and in open API for vendors to create their own unique features and functions. Having this API only available at the application layer of the OSI model allows the Zigbee standard to have interoperability between multiple devices and multiple vendors.

The Zigbee standard can operate in three frequencies: one worldwide, one outlined in Europe, and the last one available in the United States. The first worldwide frequency is 2.4 GHz. This is located in the ISM band, along with Bluetooth, 802.11, 802.11b, and 802.11g. This makes this frequency very vulnerable to interference from these other technologies or cordless phones and microwaves, which also run on this frequency. Some of the positive features of this frequency are worldwide adoption, high data rates, and the highest available channels. The Zigbee standard running in the 2.4-GHz ISM band yields 250 kbps with 16 available channels. This frequency uses a modulation technique that is different from the other frequency choices. It uses 0-QPSK. The next frequency is the one based in Europe; it is 868 MHz. This frequency has a much lower speed than if it was in the 2.4-GHz band. The current data rate on the

868-MHz Zigbee standard is 20 kbps, with only one static channel available. It runs using BPSK modulation. The last open frequency that can be used with Zigbee is the 915-MHz ISM band. This also uses the BPSK modulation, is used in the Untied States, and has a data rate of 40 kbps with 10 available channels.

## 6.8 Chapter 6 Review Questions

1. The 802.11 standard provides what maximum data rate?
   a. 11 Mbps
   b. 2 Mbps
   c. 5.5 Mbps
   d  54 Mbps

2. With Ad Hoc mode, an AP is required.
   a. True
   b. False

3. Most wireless networks operate in what mode?
   a. Ad hoc
   b. Infrastructure
   c. Repeater
   d. Bridge

4. The 802.11b standard provides what maximum data rate?
   a. 11 Mbps
   b. 54 Mbps
   c. 2 Mbps
   d. 1 Mbps

5. What part listed below radiates a wireless signal?
   a. Access point
   b. Antenna
   c. Network card
   d. Wireless laptop

6. Which of the following terms do 802.11/b/g/a have in common?
   a. CSMA\CD
   b. CSMA\CA
   c. DSSS
   d. OFDM

7. In the United States, what is the maximum EIRP limit on 802.11a UNII 1 system?
   a. 200 mW
   b. 100 mW
   c. 250 mW
   d. 300 mW

8. The 802.11a standard provides what maximum data rate?
   a. 11 Mbps
   b. 2 Mbps
   c. 48 mbps
   d. 54 mbps

9. What UNII band is for outdoor use?
   a. UNII 1
   b. UNII 2
   c. UNII 3
   d. UNII 4

10. The 802.11g standard provides what maximum data rate?
    a. 11 Mbps
    b. 48 Mbps
    c. 24 Mbps
    d. 54 Mbps

11. Which of the following PAN technologies does 802.11 address in its standard documentation?
    a. Bluetooth
    b. IR
    c. Zigbee
    d. UWB

12. What happens on an 802.11g network when an 802.11b client is injected?
    a. Nothing, the client will not operate.
    b. The access point will service both clients without any problems.
    c. The access point and all other access points that can hear the client will down-shift to 802.11b.
    d. The access point that the client is connected to will downgrade to 802.11b to support that client.

13. What standard would one need to install Wi-Fi phones in a net-work? Select the best answer.
    a. Wi-Fi
    b. VoIP
    c. 802.11e
    d. 802.11g

14. What is the newly formed draft standard that will replace 802.11g?
    a. 802.11b
    b. 802.16
    c. 802.11n
    d. 802.11z

15. Which one of these terms is not part of the wireless local area network service set?
    a. BSS
    b. IBSS
    c. ESS
    d. IESS

16. Roaming is defined in which of the following standards?
    a. 802.11
    b. 802.11g
    c. 802.11b
    d. None of the above
    e. All of the above

# Chapter 7

# Wide Area Wireless Technologies

There are many types of RF networks throughout the world. Just sitting there reading this, there are hundreds of wireless signals bouncing off your head — radio, TV, cell phones, GPS, and lots of government-controlled signals. Chapter 1 explored some of the inherent problems facing all wireless signals; this chapter looks at how companies have combated these problems in the past as well as the present. One of the biggest issues with wireless falls under the frequency spectrum. Creating how the system will work is not difficult; however, getting the OK from national governments to use certain frequency spectrums is difficult. If a non-government-licensed spectrum, such as one that is free to use, is adapted by a company, providing reliability becomes a challenge.

Security is of major concern for most of these signals. TV and radio are not very concerned about security because they do not hide their existence but rather publicize it with commercials and advertising. A lot of the signals discussed in this chapter, such as cellular phone calls, are very sensitive and require some means of security. Looking at how security was and is deployed for these different types of RF technologies gives us a better idea of how we deploy security within our WLAN (wireless local area network).

## 7.1 Cell Phone Technologies

Cell phone technology has seen three generations of change since its major commercial debut in 1980. Looking at each of the generations of technology up close, one notices the advancement of security throughout the cycle. With first-generation (G1) cell phones, we had unencrypted analog voice transmitted in plain FM. With second-generation (G2) technology, we had encrypted communication and better quality. Third-generation (G3) has us using phones to make purchases that are deducted directly from a credit card or even used to carry a line of credit with a cellular provider. Third-generation also prompted the "be connected everywhere" communication phrase. To support this requires the availability of data access technologies on mobile devices. Third-generation cellular created this data access, although it was low data rate access. This low rate access prevents advanced features.

People have already started to look at fourth-generation wireless. Using a high-speed network capable of providing the bandwidth needed to host multimedia-intense applications is just a pipe dream today. The only real solution conceivable would be a fourth-generation network made up of multiple networks using multiple technologies leveraging the fastest available delivery method. This means a wireless end device would be able to connect to multiple technologies, making decisions on what network to use based on highest bandwidth available from any carrier's networks.

One might picture a fourth-generation (G4) network as follows. As you leave your house in downtown Manhattan, your wireless device jumps from your home network to a GSM or other cellular network. Then, once in range, you might roam into one of your carrier's 802.11a/b/g/n participating hotspots or even across a longer-range 802.16e network. After you leave the congested city, your available choices as well as your bandwidth would be greatly reduced because the presence of smaller cell size, higher bandwidth networks decreases. Looking at Figure 7.1, one can see an example of how a G4 network operates.

### 7.1.1 Analog

Analog phones were the first type of cellular phones released. These phones were insecure, with voice traveling over the airwaves without any encryption. These networks were often small and suffered major interoperability problems. After the industry found this major interoperability problem, it decided to create the Advanced Mobile Phone Systems standard (AMPS). When companies started to make their networks conform to this standard, they had few interoperability problems. This standard also allowed devices from various manufacturers to work on each other's wireless networks, thus opening up the market for more competition. This competition led to more device choices and happier customers.

Looking at the security of analog cellular networks, one notes that all conversations taking place were unencrypted. This meant that with a

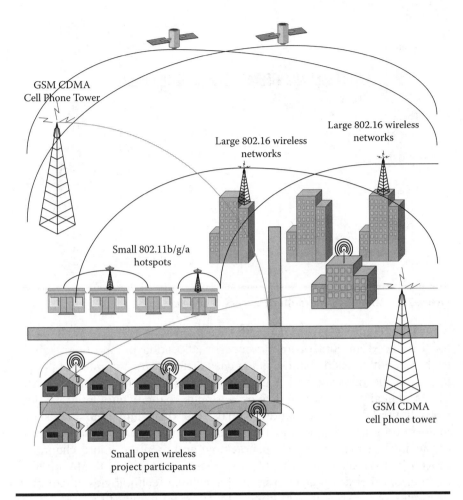

**Figure 7.1    Fourth-generation network diagram.**

simple police scanner, one could pick up and listen to anyone's calls. This became worse when criminals learned how to copy or clone these phones by extracting their identification information and copying it to a different phone.

## 7.1.2 TDMA

In the late 1980s, the wireless industry began to look at converting its analog networks to digital networks. This was to improve capacity, security, and quality. In 1989, Time Division Multiple Access (TDMA) was chosen over Motorola's Frequency Division Multiple Access (FDMA) by the Cellular Telecommunications Industry Association. This decision helped TDMA become widely used and deployed.

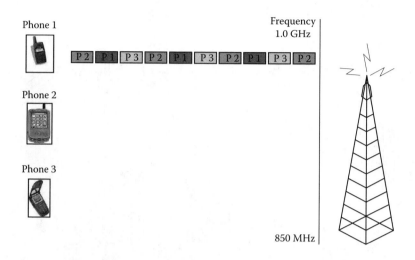

**Figure 7.2   Time division multiple access (TDMA).**

TDMA (Figure 7.2) is digital cellular transmission technology. It allows for a reduced amount of interference by allocating unique time slots to each user within each channel being used. This allows a number of people to access the network and make a call using a single RF channel. The TDMA digital transmission divides a single channel into multiple time slots. It works by taking audio signals and digitizing them into short, millisecond-long packets and then evenly spacing these inside a single channel. This takes place for a short time and then the entire channel is moved in frequency and the whole process starts over. TDMA offers a number of other advantages over analog cellular technologies. It can carry data rates of 64 kbps to 120 kbps. This allowed many other uses of this technology, such as fax, Internet access, short message service, and text messaging, as well as other multimedia application uses. TDMA also provides cell phones with extended battery life and talk time. This is due to the amount of transmitting actually taking place. A TDMA phone is only transmitting one third to one tenth of the time during any given conversation.

Of course, there are some disadvantages to TDMA technology. When roaming from one wireless cell to another, a call could be dropped. This is because there are no means of time allocation on the new cell calls. Once you roamed onto a new cell and the cell did not have an open time slot to insert your call, it was dropped. If all the time slots in the cell were already occupied when a user tried to redial the dropped call, they may not have received a dial tone.

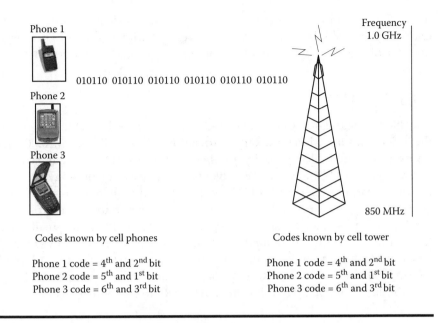

Figure 7.3   Code division multiple access (CDMA).

## 7.1.3 CDMA

Code Division Multiple Access (CDMA; Figure 7.3) was originally developed for U.S. military use in the early 1960s with the help of a company called Qualcomm. The decision to develop this technology came from a need to have communication channels more resistant to interference caused by enemy jamming devices. Also required was the need to add some form of encryption to prevent eavesdropping. Because Qualcomm had led the development of the CDMA technology, they were the owners of a large amount of intellectual property, patents, and trade secrets pertaining to CDMA. In the early 1990s, the Telecommunications Industry Association adopted CDMA as a digital access technique for cellular phones, making a once military-only technology available to the public.

This technology works by assigning a digital code to every bit of speech and then sending the encoded speech over the airwaves. All calls that are placed in the area of coverage of a single cell phone tower are placed within the same radio channel. This means that at any given time, there are numerous calls traveling the airwaves acting like a single signal. Each call is broken out from the transmission with sequential code. This code is only known to the receiving phone and the cellular tower. This code allows one to pick certain bits from the airwaves and reassemble the call, thus allowing secure communications. Defeating this code is extremely difficult because there are over 4.4 trillion codes. This large number of codes prevents eavesdropping and protects confidentiality.

### 7.1.3.1 CDMA2000

The IS-2000 standard (CDMA2000 1X) was completed and published by the Telecommunications Industry Association in 2001. The CDMA2000 1X offers twice the voice capacity of the original CDMA or what is now known as CDMAOne with the advent of this new technology. It also provides average data rates of 144 kbps. With this increased speed and data access, cellular phones can utilize their cellular connection to provide services such as Internet access, weather, news, stocks, and e-mail. This has led more and more cellular customers to converge PDA devices and cell phones into a single device. Cellular carriers have rapidly adopted CDMA2000 due to its backwards compatibility with the original CDMA and the low cost of upgrading their networks to support it.

### 7.1.3.2 CDMA 1xEV-DO and CDMA 1xEV-DV

Once CDMA2000 was available, customers quickly became interested in receiving more bandwidth. Because of this, CDMA was once again modified to increase its data speeds. This became known as CDMA2000 1xEV, which stands for 1x Evolution. Currently, CDMA2000 1xEV has two phases: (1) 1xEV-DO, which provides data only, and (2) 1xEV-DV, which provides data and voice. In 2002, 1xEV-DO systems were installed, and although they do not provide voice capabilities they are CDMA2000 1x backwards compatible. The current speeds of 1xEV-DO are in excess of 2 Mbps if the network is clear and voice calls are handled with another carrier. Because of backwards compatibility, there have been talks about having data and voice run on two different networks. This would allow allocation of maximum bandwidth to the data connection. As for 1xEV-DV, this enhancement will fix the bandwidth issue of running voice and data from the same carrier. To support these increasingly bandwidth-intensive technologies, the cellular carriers have started to converge their networks into IP (Internet Protocol).

## 7.1.4 GSM

In 1982, the Conference of European Posts and Telecommunications (CEPT) created a GSM group (*Groupe Spécial Mobile*). Its job was to define a standard mobile network for use across Europe. In 1982, GSM did not stand for Global System for Mobile Communications because it was only defined as a standard mobile network used across Europe and not the world. In 1989, the European Telecommunications Standards Institute (ETSI) took control of the GSM group and in 1990 finished the first GSM standard. One year later, GSM networks were starting to rise all over Europe. The GSM network was one of the first digital or generation-two networks at the time. Today, GSM is the most popular cellular technology in the world.

GSM itself uses FDMA and TDMA as part of its access scheme. It is quite interesting how it works; GSM takes the available bandwidth and breaks it up into 124 sub-carrier frequencies using FDMA. There are 125 available carriers, although the first one is just to protect the GSM signal from interference. The remaining 124 sub-carriers are each divided into eight time slots using TDMA. This allows four devices to use each channel. This is because each device uses two channels — one to transmit and one to receive. With this, many people have the impression that the device is simultaneously sending and receiving. Because TDMA was used to divide up the channel, there are eight time slots, which means that a device using two of the eight channels is never using them at the same time.

The setup of the GSM network is very involved; for example, the current GSM standard is more than 8000 pages long. Because of the depth of the GSM system setup, only a high-level discussion of how it works takes place here. In a GSM system, many parts are required. From a high-level view, only a select few are explored herein. The first item is the phone; most often the phone is referred to as the mobile station and designated as MS. This device holds a SIM card, which contains the phone book and all the carrier subscription information for the phone. The phone also holds the MSISDN, which is a unique identifier that informs cellular networks who the original carrier is. This is used for finding the authentication and accounting information so the carrier can send out the cellular bill.

The next device is the cellular tower. Once the phone connects to the network, the cellular tower, called a base station, receives the call. The base station then forwards the call to the base station controller. This device performs all the management of the call and relays this to the tower. This is similar to the access point and its antennas; the tower itself is only the antenna and the brains lie in the base station controller. Well before any calls take place, a number of control messages and exchanges take place. When a phone is turned on, it will try to find its location information. This can be requested by the phone or queried by the cellular network. Either way, the result is what is called a Mobile Station Roaming Number (MSRN). This number is based on location; this means that a select few of these numbers always remain in the same general physical location.

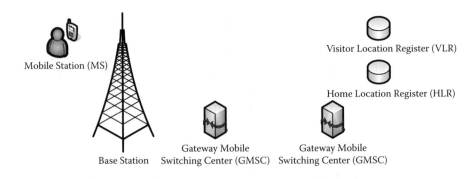

**Figure 7.4  Global system for mobile communications.**

Now, how did the phone get its MSRN? It requested it by contacting the local Gateway Mobile Switching Center (GSMC). This is a device that handles all call processing. This is very different from the functions of the base station controller, which only handles the connection management of the mobile station to the base station or cell phone to tower conversation. Once the base station controller sees the phone's request for its MSRN, it forwards the request to its local GSMC or gateway. The GSMC will use the MSIDN to find the carrier's gateway or GSMC. Once the carrier's GSMC is found, the other gateway will send it the Visitor Location Register (VLR) or one's current location and receive the Home Location Register (HLR). Now both gateways know where the device came from and where it is now. Once this process is complete, a call can take place. Figure 7.4 shows how each of these pieces interacts with each other.

The operations of a call are very similar to the control and location functions. The phone will send an "I want to call XXX-XXXX" to the base station. The base station will receive this and pass it along to its controller, which then in turn forwards it to the local gateway. The local gateway looks at the number and finds the calling number's gateway. Once the gateway is found, a request is made for the HLR and VLR so that the MSRN can be found. What this means is that the gateway that is local to the phone's current position tracks down the device, keeping track of the other mobile phone. Once the other gateway responds, it tells the current gateway where that number can be found. Once this is done, the gateway handling the call locates the other number and sends the call. As stated above, there are many other pieces to this, but this definition should help the majority of people understand how GSM works.

### 7.1.4.1 GPRS

The General Packet Radio Service (GPRS) is a high-speed mobile information service. It allows packet-based communications to take place on a GSM mobile network. This is very different from the circuit switched network that most cellular and telephone networks utilize. GPRS uses channels of GSM, or to be more specific, it uses time slots in TDMA. This allows for multiple mobile stations to use GPRS in a given area without filling up the voice channels. In these networks, voice always has priory over data. This means that in a highly concentrated cell, GPRS has low data rates. The data rate of GPRS is 170 kbps, although in real-world cases it is closer to 25 to 75 kbps. GPRS has a feature called *classes*. This allows the device to use different numbers of available channels for uploading compared to downloading. This would be relevant when the application using GPRS is one that is primarily receiving data compared to one that has equal sending and receiving. In either case, one would adjust the class feature to meet specific needs.

## 7.1.4.2 GSM Security System Overview

The security system of GSM is even more complex than the network system. This section attempts to outline it as best as possible without going so in-depth that the technical information becomes overwhelming. The GSM security model is based on a shared secret that is located on the mobile phone inside the SIM chip and the subscriber's HLR. This shared secret is a 128-bit key that is used to generate a 32-bit response. It is identified as Ki.

When a mobile station first comes online, it connects to the local gateway based on its position. This gateway then tracks down the phone home gateway and receives a random challenge and signed response from its home HLR. This also comes with two other pairs of challenge and signed responses that are used later. After receiving both pieces of information, the local gateway will only send the challenge to the mobile station. When the mobile station sees this challenge, it will use its shared secret to create a signed response and send it to the local gateway. To create this response, it will use the A3 cryptographic algorithm with the shared secret and the challenge. Once the gateway sees this, it will compare it to the already sent signed response from the subscriber carrier. Now we have taken care of authentication; this has only gotten the phone onto the network and able to place calls. The next section explores how the calls are encrypted.

Now let us return to the second pair of challenge and signed response messages. These messages were created with an algorithm other than A3. They use an algorithm called A8 to create a key that is used with another algorithm called A5 to encrypt the data. This works much the same way as authentication works. The mobile station uses the challenge and their shared secret programmed into the mobile devices to create a key. This process uses another algorithm, called A8, to create a session key called Kc. The Kc key is used with a frame number to create a unique key stream for every frame.

Another algorithm used is called COMP 128; it is used for both A3 and A8 algorithms in most GSM networks. The COMP 128 algorithm generates both the signed response and the session key in one run. The key length of COMP 128 is 54 bits instead of 64 bits, which is the length of the A5 algorithm key. Ten zeros are appended to the key as padding when generated by the COMP 128 algorithm. This means that the keyspace used to protect the key is not 64 bits, but rather 54 bits.

Over the years since GSM came out, some attacks have been released that defeat this security method. A key note that needs addressing here is that the A3, A5, and A8 algorithms are not publicly available. It has always been correct procedure for any algorithm to have open source available to prove a good level of security by subjecting it to academic and professional scrutiny. This ensures that an algorithm is actually safe. The process of having an algorithm open has been default practice for all commercial, financial, and governmental usage for more than 20 years. The GSM algorithm for A3 and A8 uses the COMP 128 algorithm as well. This algorithm was leaked to the public; and once it was securitized, it was quickly noted that a problem exists.

In 1998, members of the Smart Card Developers Association demonstrated that they could crack the A5 authentication method of GSM in a matter of hours on a single PC. This experimentation led to more in-depth analysis and later two men (Alex Biryukov and Adi Shamir) said they were able to perform an attack on A5 with a single computer in less than two minutes. Because of these threats, it is widely known that GSM systems have a poor security method.

## 7.2 GPS

The Global Positioning System (GPS) is a U.S. military space system operated by the U.S. Air Force. The system is used for position location, navigation, and precision timing. It accomplishes this using three segments: (1) satellites, (2) ground control centers, and (3) receivers. The most common consumer application is giving real-time directions to drivers; there are thousands of popular government applications, along with some special public uses such as tracking.

Satellites are the first and most widely known segment of GPS. These 24 orbiting satellites are stationed 11,000 miles above the Earth. These satellites send one-way, time-tagged transmissions that radiate down over the entire Earth. This timing is produced by four atomic clocks located on each satellite. The Department of Defense officially named the satellites NAVSTAR, which stands for *Nav*igation *S*atellite *T*iming *a*nd *R*anging. These satellites were launched in phases, with the first phase of four satellites going up in 1978 and the final phase completing full global coverage 17 years later in 1995.

The second segment of GPS is the control segment, consisting of five tracking stations located across the world. The master control center is located at Schriever Air Force Base, in Colorado. This is home to the Air Force's 50th Space Wing, the unit that provides command and control for defense warnings, along with navigational and communicational satellite tracking. The other four control centers are unmanned and controlled from the master control center. These four control centers are located in Hawaii and Kwajalein in the Pacific Ocean, Diego Garcia in the Indian Ocean, and Ascension in the Atlantic.

The location portion runs across the GPS, but all tracking services utilize some other technology to transmit the GPS coordinates. It is important to note that GPS by itself is not a tracking system. GPS receivers use the passive, one-way transmissions of signals from the orbiting satellites to determine a position fix. Nothing in the GPS infrastructure enables system operators to know who is using the signal or where they are. Conversely, a GPS unit cannot use the system to transmit its own location. For GPS to become part of a tracking system, it must be coupled with a communications device. Network-based positioning systems that use a telecommunications infrastructure can determine a user's location. This location information is either available on the device or needs another technology like cellular to transmit. A GPS-equipped digital phone can also be used to track its location and movement. These phones provide users with greater control over their own privacy because such units typically allow users to block the transmission of the phone's location information.

Looking at GPS security, one can start with its public usage debut. During President Reagan's term, GPS was made freely available worldwide for commercial use. When the system was released, it was set up to beam down two signals: (1) the Precise Positioning Service (PPS) or Precision code (PC), and (2) the Standard Positioning Service (SPS) or the Coarse Acquisition code (CA). The dual signals arose from a national security concern of allowing near-perfect positioning to enemies. This PC code is the one the military encrypts with confidential cryptographic equipment, keys, and specialized receivers. The PC code also increases its accuracy by using a second carrier wave that allows receivers to measure the small delay caused by the signals having to move through the atmosphere. The other code (the CA code) was the one released to the public. It was less accurate and easier to jam. It is easier to acquire the CA code than the

PC from the Earth; it is so easy that the military first homes in on the CA code. Once it has found and tracked the CA code, it authenticates using their cryptographic key, allowing them to transfer to the PC code. This CA code is defused by a technique called selective availability (SA). This technique allows the U.S. Government to modify the GPS signal accuracy in any place on Earth at any time. This selective availability could be used by the U.S. Government to change the course of an enemy military relying on GPS for navigation.

Although the Department of Defense tried to limit the accuracy of commercially available GPS, the GPS receiver companies soon found legal ways around selective availability. This technique, which had major involvement from the U.S. Coast Guard, is called Differential GPS. The technique worked by using the exact position of the base stations and comparing it with the base station's location through the GPS CA code. Once the level of error was determined, it could be applied to the location of the receiver, making it almost dead accurate.

Well after a number of techniques similar to the one described above were created to get around the selective availability, a major event took place. This event changed all of these concepts and techniques. On May 1, 2000, President Clinton signed an order ending selective availability and making civilian GPS readers a lot more accurate. This led to the commercial end of techniques to get around selective availability, and also created the massive proliferation of commercial GPS devices.

There have been many U.S. Government documents, reports, and appointed teams tasked with researching vulnerabilities in the GPS. Most of this work leads to the notation that the main risks in GPS are signal jamming and spoofing. Looking at what kind of jammer one would need depends on the amount of space in which one wants to prevent the GPS signals. To have any real effect on the system, one needs a jammer capable of jamming up to 10,000 watts. This would mean one would need a large truck or plane. Luckily, for the military, the bigger the jammer, the easier it is to track down and destroy. The real threat to GPS comes from a large number of small, man-portable, 500- to 1000-watt jammers. With these jammers running simultaneously and spread out in a large area, an enemy could easily create a large GPS jamming capability. For this reason, the use of small wideband jammers is of great concern to the U.S. military.

## 7.3 802.16 Air Interface Standard

The IEEE 802.16 Air Interface Standard is for fixed broadband wireless access systems employing a point-to-multipoint architecture. The initial version was developed with the goal of meeting the requirements for deployment of systems operating between 10 and 66 GHz. This has made the standard extremely large in scope, so much so that an amendment was made (called 802.16a) to finish the standard for systems operating between 2 and 11 GHz. One of the main reasons that the 802.16 standard is not taking over the DSL and cable market share is the high cost of customer premise equipment (CPE). Also in 802.16, no mobility is supported. Recently, the battle, primarily headed up by Intel, created the requirement to support mobility in the 802.16 standard. This did not make it into the 802.16 standard, although it did create the 802.16e draft standard. This could make some major changes to how people use WLAN technologies. With 802.16e, which allows laptops or other wireless devices to connect to networks that can service many miles, we may start to see pure cellular technologies disappear.

The IEEE process stops short of providing conformance statements and test specifications. To ensure interoperability between vendors competing in the same market, the WiMAX technical workgroup was created by the leaders in IEEE 802.16 technology. WiMAX is the tradename for the IEEE 802.16 standard for high-speed metropolitan area wireless networks that are interoperable. These products are expected to come to market in 2005/2006.

## 7.4 802.20 Standard

The 802.20 standard is also being created for long-range wireless access. This standard is designed to address the need to receive signals traveling at a high rate of speed, such as 150 miles per hour. This would be useful for high-speed trains. One of the ideas behind 802.20 is to have a large infrastructure where a subscriber could have an always-on connection. This would take Wi-Fi and cellular to new levels where high-speed data devices would be located in everyone's pocket and be as common as a cell phone. This standard is in the draft stage and is still very new; there is only limited information as to how it will work and at what frequency. Right now, it is looking at using an open bandwidth around the 3.5-GHz range. Frequency range is a big issue for long-range wireless solutions; using open-band, freely available spectrums means large amounts of unavoidable interference and government-regulated frequencies require a license and a hefty fee.

## 7.5 Chapter 7 Review Questions

1. What type of cellular communications is considered G1?
   a. Digital
   b. GSM
   c. AMPS
   d. CDMA

2. What type of cellular communications is considered a global standard?
   a. TDMA
   b. GSM
   c. AMPS
   d. CDMA

3. What is the data rate of the CDMA2000?
   a. 21 Mbps
   b. 144 kbps
   c. 550 kbps
   d. 2 Mbps

4. The 802.16 standard has built-in mobility.
   a. True
   b. False

5. GPS has how many satellites orbiting the earth?
   a. 24
   b. 26
   c. 28
   d. 30

6. What cellular standard was held back for many years as a military-only technology?
   a. TDMA
   b. GSM
   c. AMPS
   d. CDMA

7. Which of the following terms is the one that finds other cellular users?
   a. Gateway mobile switching center (GSMC)
   b. Home location register (HLR)
   c. Visitor location register (VLR)
   d. Mobile station roaming number (MSRN)

8. What is the documented as well as the real data rate of GPRS? (Select two)
   a. 170 kbps
   b. 210 kbps
   c. 10 kbps
   d. 55 kbps

9. What is the data rate of the 1xEV-DO?
   a. 21 Mbps
   b. 100 kbps
   c. 550 kbps
   d. 2 Mbps

10. Which cellular standard uses slices of time to allow for multiple connections?
   a. TDMA
   b. GSM
   c. AMPS
   d. CDMA

11. The A3, A8, and A5 algorithms are part of what standard?
   a. TDMA
   b. GSM
   c. AMPS
   d. CDMA

12. All the research performed on the security of GPS has led the U.S. Government to say that the biggest risk to the GPS system is _____.
   a. Security
   b. Jamming
   c. Spoofing
   d. Attacking the satellites

13. The 802.16e standard addresses mobility.
   a. True
   b. False

14. What body performs interoperability testing for 802.16?
   a. WECA
   b. Wi-Fi Alliance
   c. WiMAX
   d. Wi-FiMAX

15. What is the major difference between 802.16 and 802.20?
    a. Range
    b. Speed
    c. Ability to receive a signal at high rates of velocity
    d. Ability to securely connect to a wireless network

16. What is the data rate of TDMA?
    a. 2 Mbps
    b. 100 kbps
    c. 64 kbps
    d. 500 kbps

# Chapter 8

## Wireless Antenna Theory

All wireless devices need an antenna to send or receive signals. The antennas are what the devices use to find or send wireless signals. Without an antenna, a wireless device would have virtually no means of achieving any type of wireless communication. Placing an incorrect antenna type could dramatically reduce the usefulness of a wireless network.

This chapter familiarizes the reader with antennas by looking at how they work. Given a basic understanding of the antenna properties, this chapter then explores the different types of antennas as well as the types of connectors used. It also looks at homemade antennas — where they come from, what they are made of, and what laws exist pertaining to them.

### 8.1 RF Antenna Overview

Understanding what kinds of antennas are available as well as how they perform is a critical part of understanding wireless networking. To place antennas properly, an understanding of how they work is required. This will ensure that incorrect antenna placements do not degrade the wireless infrastructure. Given a full understanding of how antennas work and what types of antennas radiate signals in what ways, this chapter then explores which antennas are best for particular situations.

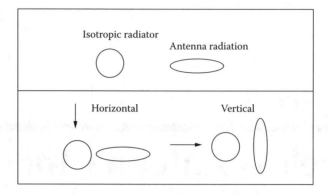

**Figure 8.1 Antenna polarization.**

## 8.1.1 Polarization

Radio waves are created by two fields: (1) electric and (2) magnetic (Figure 8.1). These two fields are perpendicular to each other, allowing energy to flow from one to the other. This energy flow is what is known as "oscillation." The position and direction of the electric field with reference to the Earth's surface determines the wave's polarization. The metal portion of the antenna that is actually radiating the signal is always parallel to the electric field. (This means that an antenna that admits an electrical field horizontally is polarized the same.) Understanding the correct polarization of an antenna can help avoid incorrectly mounting it and causing a coverage issue.

- *Horizontal polarization:* the electric field is parallel to the ground.
- *Vertical polarization:* the electric field is perpendicular to the ground.

## 8.1.2 Gain

Antenna gain is measured in dBi, which gets its value from a comparison against an isotropic radiator. An isotropic radiator is a sphere that radiates power equally in all directions, both vertically and horizontally. When we look at antenna radiation pattern against that of an isotropic radiator we see a pattern more closely related to a doughnut than a perfect sphere. When a greater gain is needed, we push two of the vertical ends together, creating a larger horizontal gain. This is similar to pushing two sides of a balloon together. When this is done, the balloon protrudes out the opposite direction that it was pushed. Both dBi and dBd are used to measure the gain on antennas. When confronted with a measurement of dBd, one can simply add 2.14 to the dBd number to produce its dBi counterpart. This is essential for an RF engineer to understand. An engineer would need to know this because, as discussed in the next section, dBi is used to find the power limitation created by the FCC called EIRP (equivalent isotropic radiated power). Most of the time, antenna manufacturers list their antennas in dBd because it is a larger number and often makes their antennas look more powerful to uneducated customers.

### 8.1.2.1 Equivalent Isotropic Radiated Power (EIRP)

A term used to measure the transmit power is called the equivalent isotropic radiated power (EIRP). EIRP represents the total effective transmit power of the radio. This includes any gain that the antenna provides, as well as any losses from the antenna cable. EIRP is the total wireless signal power output on a radio device. The EIRP value is regulated by the FCC (Federal Communications Commission), meaning that only allowed values can be used and sold. These rules and values are listed in FCC Part 15.247. All manufacturers of 802.11b/a/g products must comply with this to be able to sell their products. An RF designer must be able to measure EIRP. To do so, the following formula must be applied:

Transmitter power + Antenna gain − Antenna loss = Equivalent Isotropic Radiated
    dBm                    dBi                dB                        Power (EIRP)

EIRP is calculated by adding the transmitter power to the antenna gain and then subtracting that from any cable loss. It is expressed as transmitter power (in dBm) plus antenna gain (in dBi) minus antenna loss ( in dB). Therefore, to go through an example: a radio with a transmitting power of 20 dBm using an omni antenna, with a 5.2-dBi antenna gain connected to a low loss cable. The low loss cable is 10 feet long and injects a total loss of 5.5 dBi. To solve this: 20 + 5.2 − 5.5 = 19.7.

## 8.1.3 Beamwidth

Beamwidth is the width of an antenna's gain. When one narrows or focuses an antenna's beam, the gain will increase. There are two factors associated with beamwidth: (1) vertical and (2) horizontal. The vertical beamwidth is measured in degrees and is perpendicular to the Earth's surface. The horizontal beamwidth is measured in degrees parallel to the Earth's surface.

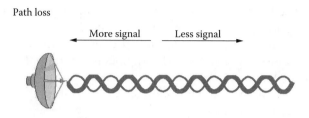

**Figure 8.2  Path loss.**

Using beamwidth when designing a wireless network is very important and one must understand what applications require a greater horizontal rather than a vertical beamwidth, or vice versa. When trying to cover a long narrow hallway, using an antenna with a greater vertical beamwidth than horizontal may prove more efficient at radiating a signal.

### 8.1.4  Path Loss

Path loss (sometimes called free space path loss) is a degradation of a wireless signal as it travels. Like all radio waves, distance often hinders the ability to keep the signal clarity at a high level. As a radio wave travels, its waveform slowly starts to deteriorate with distance. This is because path loss is occurring. Path loss is often important for bridge solutions, because they extend long distances. It is typical that line of sight stays within six to eight miles in bridging solutions to prevent path loss. If this is not followed, the path loss of the signal might prevent it from being received on the other end. Figure 8.2 reveals path loss: as the antenna signal travels, it gets degraded until it is so minute that it cannot be detected. This can be compared to the discussion in Chapter 1 relative to dropping a stone into a pool of water and watching the ripples until they cannot be seen.

### 8.1.5  Azimuth

The azimuth (Figure 8.3) of an object is the angular distance along the horizon to the location of the object. Azimuth can be used to apply direction toward the flow of a current. This current flow usually reads as a degree notation from the north point. When using azimuth with respect to wireless, its main function is to aid in the placement of antennas. We can illustrate, in a design, which way the wireless signal is propagating from the antenna by relating its azimuth. This helps orient the wireless signal's intended radiation pattern on a two-dimensional drawing.

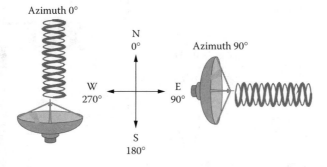

**Figure 8.3   Antenna azimuth.**

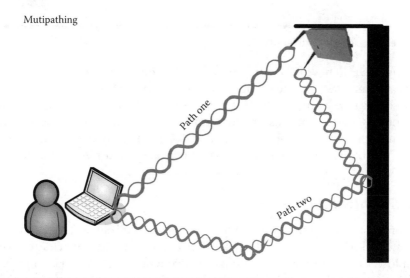

**Figure 8.4   Multipath.**

## 8.1.6 Multipath

Often, a signal will perform one of the interference types, such as reflection or diffraction. When this happens, sometimes the signal splits into multiple signals (Figure 8.4). When more than one of these signals is able to make it to the receiver, they can cause a number of problems. The reflected RF waves can travel farther than the original RF wave, thus arriving later in time than the direct RF wave. This can make it extremely difficult for the receiving antenna to put the signal back together. Another effect of multipath is distortion, which can come from having a wave that split

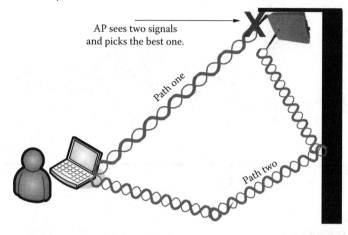

Antenna diversity

AP sees two signals
and picks the best one.

Path one

Path two

**Figure 8.5   Antenna diversity.**

during travel, creating a different wave that ended up at the receiver at the same time. When one of these events happens, it is called multipathing. This can cause issues with RF transmissions. When RF uses modulation techniques to enhance its speed, symbols are used. These symbols can become corrupt because of multipathing. When this happens, the transmitter must be able to detect that the signal is corrupted and resend the signal.

## 8.1.7 Antenna Diversity

To prevent multipathing, an antenna method called diversity (Figure 8.5) is often used. Diversity is the use of two antennas for each radio. This can lower the effect of multipathing by leveraging two antennas to receive a better signal. It does this by selecting which antenna received the best signal. This means that only one antenna is used at a time. Diversity antennas are physically separated from the radio and each other, allowing them to be placed in areas outside the propagation of multipath. Using dual antennas ensures that if one antenna is receiving a high amount of multipath and the other is not, diversity allows the radio to pick the antenna with the better-received signal. When looking to maximize the advantage of diverse antennas, make sure that they are placed two to eight feet apart. They should also be the same antenna type covering the same area. Often, people use a diversity-capable access point to cover two separate areas. This can increase the effects of interference. Also, placing antennas less than two feet apart may still help, although it will not provide maximum protection from multipathing.

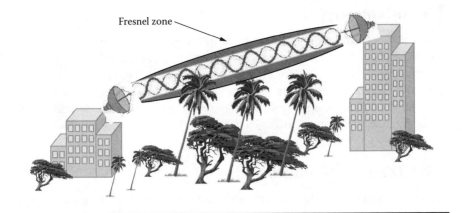

Fresnel zone

**Figure 8.6   Fresnel zone.**

## 8.2 Fresnel Zone

The Fresnel zone is the area around a wireless signal that needs to be clear of any obstructions in order for a wireless signal to be accurately received. The wireless signal itself is not a small airwave; it has a large breadth and objects can often get in the way. When looking for a line of sight, which is needed for wireless bridges, one must take into account the Fresnel zone.

Looking at Figure 8.6 you can see the fresnel zone affects long range communication. As one can see from the figure an object can interference the fresnel zone and this can create communications issues while still having good line of sight. To fix this issue, a large tower is needed to get the antenna higher than this Fresnel zone.

The Fresnel zone is also impacted by the curvature of the Earth. As Christopher Columbus discovered, the Earth is round. How he proved this teaches us about how a wireless signal could become blocked simply by the curvature of the Earth. When Columbus told everyone that the world was round, the public asked him to prove it. When he left on a short trip out to sea, he asked everyone to watch his sail as he traveled. As he went out to sea, his mast started to slip out of sight. As the people on shore watched, right before his ship went totally out of sight, all they could see was the tip of his sail. Columbus said, "If the world was flat you would be able to see the entire ship"; but because the world is round, the body of his ship slipped below the curvature of the Earth. Funny enough, no one really bought this theory and he still attempted to travel around the world to prove his point.

# 8.3 Antenna Types

Given a basic understanding of how antennas are measured and what variables affect their usage, one can start to look at the different types of antennas and how each antenna type propagates signals in a different patterns. Given a full understanding of the different types of antennas, one can then look at some of the primary connectors used. Finally, there are homemade antennas similar to the ones primarily used by war drivers.

## 8.3.1 Directional Antennas

A directional antenna has a radiation pattern crafted to flow in a single direction. This single direction flow is very difficult to accomplish because all directional antennas will radiate a small amount of signal 360° around the antenna itself. If one were to look at how a directional antenna is created, one would see that when one squeezes an isotropic radiator, one pushes the signal out one direction while some of the signal squeezes out the other end. This means that if one is standing directly behind any directional antenna, one will see that the signal still is able to radiate the opposite way of its intended direction. When looking at antenna placement, directional antennas are best used to shoot signals down long narrow paths such as hallways and aisles. They are also good for short-range bridging applications. Visit a large warehouse such as Home Depot and one can see the usage of directional antennas. The most common directional antenna is the Yagi. This antenna generally has a 17° beamwidth, closely representing a triangular cone shape.

In general, the parabolic dish is specifically used for long-range bridging applications. Its beamwidth and rotation pattern are very similar to that of the Yagi except the signal will travel much farther. The patch antenna is more of a middle ground between directional and omni-directional, as its beamwidth is 180°. The patch antenna is often used against walls because of its 180° coverage pattern. It is still classified as a directional antenna. Another reason the patch is often put against walls is to prevent the wireless signal from leaking outside. Figure 8.7 shows both Yagi and patch antennas.

## 8.3.2 Omni-Directional Antennas

Looking at other forms of antenna, there are the omni-directional antennas (Figure 8.8). What differentiates omni-directional and directional antennas are the beamwidths or coverage patterns. All omni-directional antennas have a 360° coverage pattern. This means that they radiate equally in all directions. Omni-directional antennas are best suited for large, open areas

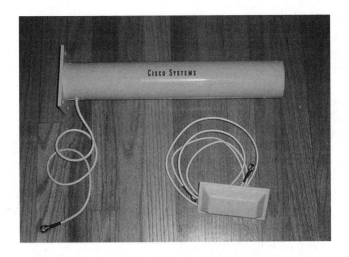

**Figure 8.7    Directional Yagi and patch antennas.**

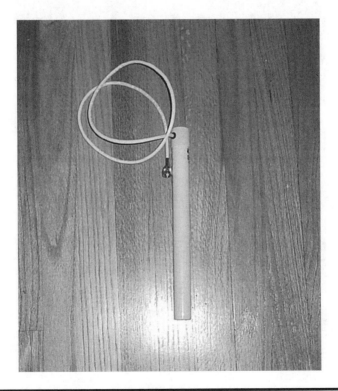

**Figure 8.8    Omni-directional antenna.**

that need coverage. Placing omni-directional antennas against any walls or directly on a column will prevent their ability to properly radiate the signal. Often, omni-directional antennas are hung up against columns; this is due to their usage in large, open areas such as warehouses. When this happens, it is very common for someone to mount the antennas right against the column and not place them out from the column where they would operate more efficiently. To get the best coverage, one should place omni antennas at least three feet from any column.

### 8.3.3 Homemade Antennas

Homemade antennas (Figure 8.9) have become popular due to the large price tag on commercial antennas. A Yagi antenna, similar to the one constructed out of a simple Pringles can, will sell for almost $200. There

**Figure 8.9 Homemade antenna.**

are even some Web sites detailing how to make a Pringles can antenna — they say the cost is about $6.50. One thing to note here is that the FCC considers homemade antennas illegal. There are some homemade antennas for sale. Outside the United States where the FCC has no authority, one can purchase homemade antennas. One place in Canada operating an online shop (http://www.cantenna.com) is selling homemade Pringles can antennas.

## 8.4 Connectors

Connectors are used to hold the antennas to devices. They can be used to hold the antennas on access points or wireless network adapters. There is not much available information associated with antenna connectors. However, one needs to ensure that one uses or orders the same type of connectors or the antenna will not fit. Wi-Fi antennas are very different between access point manufacturers. This means that access point vendors need their own connector types. This section identifies the most common connector types, looking at what access points or wireless network cards use each type. The reasoning behind why there are these defined antenna types goes back to the governing body, the FCC. The FCC wanted to prevent people from using common high-gain antennas and affixing them to wireless access points, thus breaking their predefined power output. This is why they made the industry use those antenna types that are not common across other types of antennas. This is referred to as the "use something uncommon and you will be OK" rule.

### 8.4.1 N Connectors

The N connector was created by Paul Neill of Bell Labs in the 1940s. It was named after Neill and call N for short. The N connector is common among antennas; most antennas come with N connectors. This is a little different from Wi-Fi connectors. Almost no access points use N connectors to attach to antennas directly. When ordering antennas, some only come with the N connector and a converter is needed. Knowing what the N connector looks like will help prevent problems. Figure 8.10 shows what an N connector looks like.

### 8.4.2 Reverse-Polarity TNC-Type Connector (RP-TNC)

The reverse-polarity TNC-type connectors (Figure 8.11) are one of the most common access point connectors. This connector, like the N con-

**Figure 8.10   N connector.**

**Figure 8.11   Reverse-polarity TNC-type connectors.**

nector, was created by Paul Neill, although in this case another person also contributed, giving us the Concelman part of the name. His name is Carl Concelman and he worked for Amphenol. In the late 1950s, Neill and Concelman created the TNC connector to reduce electrical noise, which was causing problems in the common BNC-type connectors. This noise was only seen under extreme vibrations. The reverse polarity part

(RP) is created by a keying system with a reverse interface. This ensures that reverse polarity interface connectors do not mate with standard interface connectors. Concelman and Neill accomplished this by inserting female contacts into plugs and male contacts into the jacks. Some manufacturers might use reverse threading to accomplish reverse polarity. Either way it is done, it meets the mandate set forth by the FCC that each wireless LAN antenna type be an uncommon one. To meet this criterion, many manufacturers used this antenna and reversed the polarity to create the RP-TNC connector.

A large amount of access points, including Cisco, Linksys, Proxim, and Sonicwall, use the RP-TNC connector. For the most part, the access point is the female side of the connector. This means that the antennas connecting to these units must have male RP-TNC connectors. Most antennas sold solely for 2.4-GHz wireless systems have male RP-TNC connectors. If the connector is an N connector as described above, then a pigtail converter is needed. This allows the RP-TNC connector to work with an antenna with an N connector.

### 8.4.3 SMA, RP-SMA, and RSMA

All of these connectors connect a client card to an antenna. In SMA (subminiature version A) connectors, as for RP-TNC, each manufacturer met the odd requirement by using reverse polarity, thus giving the RP portion. The SMA connectors were created in the 1960s. The connectors in this section primarily are used for Peripheral Component Interconnect (PCI) wireless cards. This means that they are used in desktop computers, as compared to Personal Computer Memory Card International Association (PCMCIA) cards, which work inside laptops. The SMA, RP-SMA, and RSMA connectors are used to connect the wireless adapter to an external antenna. When a wireless card is used on a desktop, the placement of the tower and the antenna is most likely behind the unit under a desk. This is not the best place to put an antenna; because of this, these connectors allow for a variety of antenna types and the ability to place them on top of the desk where they are more likely to get better signal levels.

### 8.4.4 MC and MMX

These types of antenna connectors are mostly found in PCMCIA wireless cards. MMX is an acronym for "micro miniature"; these connectors are called that because of their small size. The MMX connector type was created in 1990. Most wireless cards fall into this category. Usually, these connectors are placed with a fixed antenna on the card, making the cable

**Figure 8.12   MC and MMX.**

never seen. Some cards allow for antenna modifications. To allow for adding an extra antenna, they either offer no antenna and one that allows for the connectors, or a fixed antenna with one of these connectors that bypass a fixed antenna. Looking at Figure 8.12, one can see that one of the cards has no antenna and just the connector, and the other one has the connector at the end of an antenna. On the second one, when an antenna is hooked up to it, it will bypass the fixed antenna and use the external antenna. Knowing which one of these connectors is on a radio card will ensure that the correct antenna is ordered. Figure 8.12 shows these types of connectors on a couple of PCMCIA wireless network cards. These connectors are some of the most common types of pigtails used for war driving. When someone wants to war drive, he or she will most likely find a way to use powerful antennas on their client cards. To note is that sometimes, overpowering antennas like the Yagi can actually reduce the strength of an antenna when connecting to a client card. This is due to the small amount of power outputted from these cards compared to the large amount of power needed for a large Yagi antenna.

# 8.5 Chapter 8 Review Questions

1. When radiating an RF signal to create an area of coverage, what part of the WLAN is used?
   a. Client card
   b. Access point
   c. Antenna
   d. None of the above

2. What happens to the radiation pattern of an antenna when the gain is increased?
   a. The angle of radiation becomes larger.
   b. The angle of radiation becomes smaller.
   c. Nothing.
   d. The coverage decreases.

3. Which antenna allows for coverage in a 180° area?
   a. Yagi
   b. Patch
   c. Omni-directional
   d. Directional

4. What happens when a 50-foot antenna cable must be replaced with a 100-foot cable?
   a. Nothing.
   b. The area of coverage will decrease.
   c. The area of coverage will increase.
   d. The antenna cable will reduce the access point's transmission rate.

5. Which antenna allows for coverage in a 360° area?
   a. Yagi
   b. Patch
   c. Omni
   d. Directional

6. EIRP is a representation of the power that an entire RF system has. When correctly measuring EIRP, what should be measured?
   a. Cabling output
   b. Radio output
   c. Antenna output
   d. Client card antenna input

7. What connector is used on a 2.4-GHz 1200 Series Cisco access point?
   a. RP-SMA
   b. SMA
   c. RP-TNC
   d. MC-MMX

8. How does one calculate EIRP?
   a. Antenna gain + Antenna loss
   b. Antenna gain + Cable loss − Transmitter power
   c. Transmitter power + Antenna gain − Cable loss
   d. Cable loss − Antenna gain + Transmitter power

9. To help with null areas and multipathing, what antenna architecture would you recommend?
   a. Splitters
   b. High gain antennas
   c. Low loss cables
   d. Diversity

10. Some access points have two antennas on them. Why? (Select two)
    a. To increase the speed of the wireless
    b. To reduce the effects of multipathing
    c. To provide a hot standby antenna
    d. To provide antenna diversity

11. There is a customer who wants to connect one office building to four warehouses in the area. Most of the offices are located less than a block away; the farthest one is half a mile away. Placing what antenna on the roof of the office will provide the best coverage?
    a. 13.5-dBi Yagi
    b. 6.5-dBi patch
    c. 21.5-dBi dish
    d. 5.2-dBi omni

12. What is the designation for dBi? (Select two)
    a. Dipole
    b. Decibel
    c. Baud
    d. Isotropic

13. The area around or immediately surrounding a wireless signal between two locations is known as the _____.
    a. Free space loss
    b. Line of sight
    c. Fresnel zone
    d. Signal area

14. What is the typical line of sight when using wireless between two buildings?
    a. 2 miles
    b. 11 miles
    c. 15 miles
    d. 6 miles

15. You are trying to provide coverage down a long hallway. What is the best antenna solution for this?
    a. Yagi
    b. Patch
    c. Omni
    d. Directional

16. Determine the EIRP for the following configuration: a 20-dBm radio using a 13.5 Yagi antenna with a 100-foot cable at a loss of 8 dBi per 50 feet.
    a. 17.5
    b. 254
    c. 33.5
    d. 29.5

17. You are connecting two buildings together with a wireless bridge. You are using a Cisco bridge with a Yagi antenna. In between the two buildings are some trees. You are unable to correctly set up a connection. What should you do?
    a. Call Cisco TAC.
    b. Use a dish antenna.
    c. Raise the antenna above the treetops.
    d. Use a high power setting.

18. If an antenna has a 6-dBd rating, what is that rating in dBi?
    a. 8.12
    b. 9.31
    c. 4.8
    d. 8.14

19. When an antenna gain increases, its beamwidth _____.
    a. Increases
    b. Decreases
    c. Stays the same
    d. Multiplies

20. Which antenna is the most directional?
    a. Yagi
    b. Patch
    c. Omni
    d. Parabolic dish

21. What is the designator for dBd?
    a. Dipole
    b. Decibel
    c. Diversity
    d. Baud

# Chapter 9

# The Wireless Deployment Process

The process of installing a wireless network is often simple for a single deployment. When looking at a controlled process that is repeatable across the entire enterprise, there are number of steps that one must address. This chapter explores each of these steps, looking at ways to implement controls that help lower costs and create consistency across the enterprise. When deploying wireless networks on a large scale, one must look at the most controlled, cost-effective, and repeatable process available. Getting to this requires some up-front work and a lot of experience.

When deciding to take on a large wireless project that spans the enterprise, one of the first decisions is whether to perform the work in-house or outsource it. Often, using an outsourced model is one of the better ways of tackling a wireless deployment. When vendors are involved, some of the risk is shifted from the organization to their vendors themselves. Ultimately, the company is still responsible no matter what. If this process is properly handled, the vendors assume any additional financial risk outside the contracted amount. Often, unforeseen financial burdens will arise, stemming from a poor design, unskilled designer, malfunctioning hardware, or any other technical hurdle that might impede the timing or cost of a project. This approach is more difficult, if not impossible, to apply to smaller organizations; however, for larger companies, this route performs especially well. When upper management is more financially driven than technically driven, having the ability to fix the cost associated to any technological deployment will greatly improve your worth as an employee.

Looking at the entire deployment process, we can walk through a fictional organization (MoreCo Corporation), addressing its requirements to deploy wireless. The wireless system will be installed to support the organization's physical security force. MoreCo needs to deploy wireless at 20 locations nationwide. In starting this process, one gets into the requirements for wireless, looking at the need for wireless. This section addresses the why and what the organization can expect to gain from this technology. This will help ensure that at the end of the day, the project requirements allow the customer to achieve what they originally needed. In walking through the entire process, one will see how MoreCo handled each step of the process. This information will help by providing a more true-to-life example of how to perform the steps outlined in this chapter.

If a deployment falls within the size scope described above, a key element would be to have an independent, technically qualified consultant available. This consultant might be an internal employee. If an internal employee is used for this role, make sure he or she is technically qualified and has sufficient experience within the wireless industry. One of the best approaches is to hire a short-term consultant for the duration of the deployment.

The focus of this consultant would be to oversee all technical aspects of the process. Having this resource involved can bring valuable insight to your team and can help develop and implement a number of controls identifying any technical shortcomings of the various vendors. By using this resource in an independent auditing role, one can often find several issues that otherwise might not be brought to anyone's attention until it is too late. Having this kind of expert analyzing all of the designs and installations forces the vendors to optimize designs and install the most technically applicable wireless networks possible.

## 9.1 Gather Requirements

First, one must look at exactly what a company will use its wireless network to accomplish. This is one of the simplest tasks for many people because the greatest benefit of wireless is the increase in worker productivity. There are many other applications for wireless across many vertical markets, such as retail, manufacturing, education, healthcare, and many others. In these industries, one often sees wireless being used as a tracking method to reduce the overhead and labor costs associated with inventory. In addition, one sees wireless point-of-sale systems appearing in the retail market, allowing easily movable storefronts. In healthcare, one sees doctors and nurses using wireless tablets, creating a virtual clipboard with all the hospital records on it. There are a number of reasons in each vertical market for deploying wireless. The main goal here is to identify the driving requirement for wireless.

After determining what the driving requirement or requirements are, the next step would be to find out how much wireless is needed and where. What this means is, do you need wireless everywhere within your facility or maybe just a small area? Understanding the reasoning behind deploying wireless gives a good idea of where within a facility or campus it is needed. In a wireless deployment, there are often many people involved with many different skill sets. Gathering requirements from each group, as well as understanding them, is a key role in the requirements stage of a project. A person who is not technical may have requirements that unbeknownst to them may need technical solutions or may change the general requirements altogether. This is often seen when someone requests wireless access inside places like a cooler. It is possible to place wireless inside a cooler, although it will change the pricing. This pricing change may be substantial enough to drop that requirement altogether.

In applying this section to the example company MoreCo, one can see that the businesses need is to wirelessly hook the physical security force into the camera system from their PDA devices. This means they need wireless in all the areas of the facilities that the security force are required to guard. This would be a prime case where understanding the skill sets of individual job roles would help to make sure that the most cost-effective view is taken. At MoreCo, there is a shed at one location; this shed is more than 2000 feet from any telecommunications closet. To add to this, there is no line of sight to use a bridge solution. What this means is that if wireless is needed inside the shed, a major cost increase will occur because fiber optic cabling will be needed to support that single access point. This cost increase may be fine with the technical team leaders; however, the managerial team leaders may decide that the cost is too great and change the requirements away from covering all buildings. The next requirement to address is the bandwidth required to support video. Video is real-time, which means it needs bandwidth. This means that certain data rates must be everywhere there is coverage. This is needed as a requirement so that when the wireless network is put in, it can support the high data rates required to stream video.

Having dealt with the main requirements issues, one now understands why wireless is needed and where exactly it is required. One also knows that the wireless network needs to support time-critical data exchange stemming from video. Other key points that one must address include when it is needed, how much it will cost, and how it should be done.

## 9.2 Estimation

With a good understanding of what a wireless network will be used for and what the requirements are, one can then look at one of the most significant factors: the cost. One needs to gather this prudent data to move on to the next step, that of making a business case. In looking at the cost of wireless, many factors create a fluctuating cost. Even in buildings that are designed by the same architect and have exactly the same layout, one will see a difference in RF design. This is due to the many environmental variables that are outside one's control. These variables often influence wireless coverage patterns, changing the amount and placement of equipment.

The *estimation* stage is listed second in the process. This is done for a two main reasons. First, to get an accurate cost of a wireless deployment, one must perform a site survey. This site survey takes time and resources to perform and, because of this, the survey often has a cost associated with it. Second, if one were to make a business case before actually getting an estimated cost, one could not properly report one of the biggest factors needed in the decision-making process. Therefore, one needs to have a sound way to estimate cost without performing any cost-prohibitive tasks.

The estimation process of a wireless network can be a tricky one for a number of reasons. The best way to approach this is to use an average area of coverage based on a single access point and then divide that number by the total square footage of the area where coverage is required. Keep in mind that this will change, depending on the technology used (802.11b/a/g). This is because each technology has its own area of coverage size. Right here would be one of the first areas that independent consultants could come into play to provide personal advice and experience on the subject matter. They will know, based on the requirements, what kind of technology will best suit everyone's needs. This will give a rough estimation of cost. Although it will not be 100 percent accurate, it should prove close enough to make the business case and receive the funding needed to proceed to the next step.

Many other costs must also come into effect during this stage. The wired network must be in place and able to handle the access points. There might be cable length issues and other issues that may incur extra costs (e.g., outdated equipment). Most of this detail would not be discovered until the designing phase. If the correct questions are not posed upfront during the estimation stage, the cost could change dramatically. This would be another reason for having an expert on your team who knows what has burned them before and now knows what to avoid.

MoreCo made a choice to use the 802.11g standard. They made this choice based on testing a small consecration of access points from the vendor that they wanted to use. This testing proved that their application functioned correctly and handled all the requirements correctly. They then took this number and applied it to the formula above to estimate the number of access points they might need. Afterward, they sent checklists to all 20 sites. These checklists were created by the independent consultant, outlining questions that if answered incorrectly would raise red flags. One example would be: "Do you have locations requiring wireless coverage that are more than 300 feet from a telecommunications closet?" Once these were sent and returned, an estimate can be created. This estimate provides a number close to what the project will actually cost.

## 9.3  Make the Business Case

This section looks at what is required to convince management that a wireless network is necessary. Using the increase in productivity is often very helpful in this area. When looking at the other vertical markets, inventory tracking, material handling, and general mobility often make good business cases of their own. In creating this business case, using real-world statistics from reputable marketing firms will reinforce one's business case. If using a wireless network to increase productivity, try using some of the research done on the amount of increased productivity due to a wireless LAN. This has been researched and documented by a number of marketing firms.

One of the key points in making a business case is return on investment (ROI). How can this wireless system save the organization money? Is it by reduced costs for material handling or labor associated with inventory tracking? Is it increasing productivity because team members will be connected no matter where they are inside their office or campus environment? There are many benefits to wireless networking; to make it easy to quantify cost savings into ROI, one must understand the benefits from a cost-savings or profit-producing prospective. This is one of the more challenging tasks. Once one creates numbers based on anticipated savings or anticipated productivity gains, those numbers easily become speculation. As pointed out above, the best way to approach this is to use industry marketing information as the basis for the amount of increased productivity. Most management teams that do are only interested in technology that can improve their bottom line and not introduce excessive costs to support and maintain. If one can present a wireless network that saves money and is backed up with statistics from other companies, then one's business case will soon become a funded project.

When looking at the business case from the standpoint of MoreCo, one can see that the benefits include security and cost reduction. Because of these benefits, the new wireless system will improve incident response time. MoreCo will see a benefit of better security because their physical security team members will be in constant contact with all available cameras and command centers. One can see the benefit of cost reduction because now when team members respond to an incident, they are not required to have someone watching the camera and reporting what they see. With the new solution, the responder can get updates from his PDA en route. This solution saves money associated with having two security personnel responding to each dispatch.

## 9.4 Site Survey

Once management has signed off and funding has been acquired, the first step of the physical work takes place. The site survey is used not only to get an accurate cost, but also to get the correct data needed to design the wireless network properly. When performing a site survey, there are a couple of requirements. The first requirement is equipment. The equipment needed to perform a site survey consists of what is called a site survey kit. Inside this site survey kit is an access point, a large array of antennas, a portable battery pack, and a large vertical boom used for raising the antennas. This equipment is used to try a number of different antennas to achieve maximum coverage within any given area. The surveyor will need architectural drawings showing the layout of a building or campus. The surveyor can create these drawings, although it is more cost-effective to provide them. The site survey team also needs to have a spectrum analyzer so they can locate and identify any interference. Depending on whether a bridge solution or a large outdoor solution is needed, the site survey vendor might need a GPS device.

This section discusses how to perform a site survey and how to properly manage it to the customer's advantage. Many companies perform site surveys. Choosing the correct company requires a little background work and proper vendor management. The first process is to obtain a qualified and reputable site survey vendor. To do this, one would need to release a Request for Proposals (RFP) or a Request for Quotes (RFQ). To maximize the success level, there should be a number of vendor requirements inside this document. These requirements are one area of the process where one can implement some technical as well as financial controls to mitigate risk.

### 9.4.1 Performing the Site Survey

When performing a site survey, there are a number of items to address. Before one can perform the survey, one needs to make sure that one has the correct equipment. Bringing the proper equipment will ensure that nothing is missed and that all the surveys are performed correctly. One thing to note about equipment: if performing any type of work far away from one's office, always assume that everything will break. With this said, if you are performing the survey and something breaks, it is your fault for not assuming that everything will break. If managing a wireless project and a team member or vendor says that something broke and it prohibits them from working, make them aware that it is their fault. They should have assumed that everything will break, and not doing so is a lack of experience on their part.

Many companies will sell full-blown kits that have most of the items discussed above. In every kit, there will most likely be something that is needed or not provided. This subsection looks at each of the needed items, identifying why one needs them and how they work. Then it examines the process of performing the survey.

The first item one needs is an access point. This access point must be the same model that is going to be installed after the survey. Using a different access point model for a survey from the one that will be installed can and often does create coverage problems. After performing the steps outlined in the gather requirements step, one should choose a wireless transmission technology. Now all that one needs is direction on the access point manufacturer. Currently, there are several manufacturers. Making a choice between them depends on a number of factors, such as current relationships, industry leaders, special requirements, features, or any other factor. Some of the more dominant access point vendors include:

- Cisco
- Aruba
- Proxim
- Symbol
- SonicWall
- Netgear
- Linksys
- D-Link

The next item is a method to power the access point. Providing power to the unit is critical. Not only does it need power to operate, but it also needs power without dragging cords around the building being surveyed. In most cases, there will not be a plug available to use where one wants to place the access point. To fix this, one can purchase a battery to provide the access point power. Depending on the length of the survey, one battery can last for the duration of the survey (two batteries may be required). One should always come prepared, always come expecting everything to break. Some companies sell these batteries as part of their surveying kits. Use the Internet or access point manufacturer to locate companies that sell site survey kits.

The next item is a telescoping pole, which is needed to hold together the access point, antennas, and cabling. This pole must have the ability to be raised and lowered to accommodate the height at which the access point will be mounted. Normally, this pole is constructed or purchased with the surveying kit but some kits do not provide the pole. The pole should be light and telescopic so it can be broken down to be shipped or put on a plane. The pole should be a bright color to prevent people from running into it. This may sound like a small issue that can be skipped — until a surveying team is told it cannot erect the pole due to it not being a highly visible color. Many safety regulations require this, so make sure this does not become a showstopper.

To perform a survey, an array of antennas should be part of the surveying kit. These antennas allow the surveyor to have maximum flexibility when placing access points. There are many types of antennas for access points. Using all of them is more of a hassle than a benefit; not only does this become a problem for the surveyor to lug around, but it also becomes a problem in supplying spares. When a network has nine or ten different antenna types, keeping all the spares can waste some money, especially if two of the antennas are very similar. The best way to handle this is to standardize on a couple of antenna types. To figure out which ones to pick and which ones not to pick, one should look closely at which antennas are the same type and close in power rating. Choosing two patch antennas, one at 5.5 dBi and one at 5.2 dBi, is a clear-cut example. Having the same type of antenna with less than 0.3 dBi difference is a waste of money and time. Also, when choosing to limit the allowable antennas, make sure to include, at a minimum, one of each type. An example of this would be to choose one patch, one omni, one Yagi, and ceiling-mount variation. This will ensure that one is providing maximum flexibility without increasing complexity.

Another, hopefully obvious item is a wireless end device. Most likely, this will be a laptop with a wireless card. The wireless card should be the same model that will be used in the end clients. This is not always possible, although it is recommended. In addition, when looking to pick a wireless card for surveying, make sure to not have an extremely high- or low-powered card. Using an average card will ensure that the survey is as accurate as possible. Two of the same cards from the same vendor can often have very different signal results. To make sure the survey is as accurate as possible, use a number of cards beforehand to find the most common one. Having found a common card not too strong and not too weak, mark that as a surveying card and use it. If one uses this same card for awhile, perform the testing again and always make sure that at least 25 percent of the cards used for testing are ones that have not been tested previously. When making an initial surveying card purchase, make sure not to buy the cards together. The differencing factors of client cards can rarely be seen from cards that were part of the same factory run.

Some other key items to bring include the tools required to document the results. These tools include a set of drawings, colored pencils, bright tape, and some colored sticky dots. The pencils and dots are used to draw the coverage patterns and corresponding access points on the drawings. The bright tape is used to indicate the actual area where the access point should be placed. The tape is affixed to the column in the exact place that the access point will be hung.

The final item is a spectrum analyzer, which is used to find any interfering signals that might impede a wireless network. Because wireless local area networks work in unlicensed airspace, other devices might also be using the same frequency. In addition, other access points might interfere with the survey. These access points could be from other, nearby companies or employees who could not wait for wireless. One of the reasons for the analysis is to help the customer and the surveyor know that the air at the frequency which the wireless network is operating in is clean. Many types of spectrum analyzers are available, ranging from inexpensive $1000 models to high-end, $100,000 models. To perform a simple analysis to see if another signal exists is easily done with the least expensive models. To use the analyzer to troubleshoot problems, a high-end analyzer is a more appropriate choice. Note that the higher-priced analyzers are available to rent. This means that in the rare event that one might need one, there is no need to worry about shelling out $100,000.

To perform the survey, a team usually consisting of two or more people will sit down and get a good idea of where and how they want to cover a given area. All too often, the surveying team goes out and just starts putting up the pole and walking off the coverage circles without sitting down and planning the logical placement of the access points. When this happens, there is often a waste of coverage and a crude-looking design. The correct way to perform a survey is to perform a draft run on paper, looking at the most feasible antenna options for particular areas. Understanding this and following it will improve time and reduce the cost. Once this is completed, the survey physically starts.

Now the team goes out to the first location and raises the telescoping boom with the access point and the first antenna candidate. Once this is up, they connect to it with the laptop and make sure that the signal looks good and the device is sending and receiving traffic. Next, they walk off the area of coverage. This is done by locating the edge of the signal all the way around the access point. There are many tools available to do this. Chapter 1 discussed SNR values; these values play a key role in determining the end of the cell's coverage. Once the end is found all the way around an access point, it is detailed on a drawing with a colored pencil and a dot is placed where the access point should go. The next step is to mark, using the bright tape, the location of where the access point should hang. Affix the bright tape to the column or something that represents where the access point will be mounted. After this, the team moves to the next location and repeats this process. The entire area where the customer asked for coverage goes through this process.

After performing the survey, the drawings must be converted to an electronic format. The most common format is auto-CAD. During this process to convert the documentation, channel selection must be addressed and designed. When performing the survey, looking at a channel structure is omitted. One is not required to pick channels during the site survey. Performing a spectrum analysis makes sure that no interference is inherent on any of the available channels. Because of this, it is common for a site survey designer to develop a channel structure when all the access points are listed on the drawings. The next set of documentation involves a physical information sheet showing all the access point locations.

Inside the physical information document are the following items:

- Location of access point
- Placement of antenna
- Type of antenna
- Power settings
- Channel settings

Going deeper into the physical information document, take a look at each of these items, what they are, and why they are needed. The *location* portion documents the bay and column where the access point is supposed to be mounted. This helps whoever might install the access point find the correct location. During an access point installation, placement of the access points is very critical. Moving the access point more than seven feet from where it was surveyed can dramatically affect its coverage pattern. This is why location information is required in the site survey design.

The next two items, *antenna placement* and *antenna type,* are addressed together. These are used to detail and explain how and what antennas are mounted and in what locations. This is important because certain antennas must be mounted in certain ways. Mounting a patch antenna pointing the wrong way will quickly become an issue. With correct details, this can be avoided. Having this item in the documentation ensures that events like this never happen. This also helps against installing the incorrect antenna altogether.

The next portion is very important; it details the access point's *channel and power settings*. The power settings dictate how big a coverage pattern is. Most commonly, a site survey vendor will blast these devices with full power. This is not a recommended practice because one does not have the ability to increase the power. The best practice approach is to inform the site surveyor that they cannot use the maximum or minimum power settings. This allows for flexibility if a coverage issue arises in the future.

## 9.4.2 Technical Controls

As discussed in Section 9.4, there are some controls and requirements that can be put in place. When using a company to perform the survey, a number of requirements and controls must be put in place well before any work starts. If these controls are thought out and implemented in an early stage, the risk involved will be dramatically reduced.

One of the first controls would be to ensure technical competency. Having a statement requiring the vendor to have trained and certified individuals on staff is a start. All too many times, companies win bids for project they subcontract out. When they do this, they do not require the same technical competencies originally asked of them. To make matters worse, depending on how it is worded, they may have fulfilled the requirements just by employing technically competent resources. This does not mean they will work on your project; be careful of this, as it happens all the time. Making sure to stipulate that certified individuals are working on your project is one of those small details that can create big problems later if not addressed up front.

## 9.4.3 Financial Controls

Looking at a financial control, first make sure that the requirements for the vendor are clearly defined and specifically detailed. This must be complete and accurate if this control is to work correctly. Understanding exactly where and what must be covered by a wireless signal is necessary for this control to work. If any of this information is incorrect, the control will lose its power and become meaningless. If all this is done correctly, the agreement between customer and site survey vendor can hold the site survey vendor responsible for ensuring that the survey will adequately cover the required area until certification.

Providing warranties for a site survey has long been a gray area for many site survey vendors. This is because environmental variables can adversely affect a site survey. This means that if there is no interference when a survey is performed and then suddenly there is, who is responsible? This again is when and where an independent party can come into play, providing positive feedback from previous experiences. Most site survey vendors will only guarantee their surveys for a very limited time. Every day the site survey is under warranty past the day the survey was performed becomes additional risk for the site survey vendor. This is why site survey vendors have such short warranties. Correctly weighing this risk versus the project timeline is a tricky task that must be adequately understood by all involved parties.

Correctly documenting the length of time and full extent of contractual obligations to warranty a site survey is necessary before any work begins. If this step is done correctly, the site survey contractor is fully responsible for the financial cost of designing, installing, and certifying any additional access points if the required area was not fully covered. This can also go the other way with too much coverage; too much coverage will cause just as bad of problems as not having enough. Having too much coverage is extremely difficult to control from a contractual standpoint. Having the consultant make his recommendation is key at this point. If the survey looks excessive in relation to the total number of access points, the consultant can easily point that out.

When we look at MoreCo specifically, their best option would be to have the independent party write up a Request of Quote (RFQ) document. This RFQ should detail the length of warranty and the extent of what financial responsibility the vendor would have in the event of a coverage issue while under warranty. The control on too much coverage is implemented by having the consultant approve the site survey document before it can progress to the next step of design. More then likely, this party would have close interactions with many vendors that perform this service and would already have a good idea of who has successful and unsuccessful surveys under their belt. This could become an added benefit by allowing other specialized companies that the consultant was aware of the opportunity to bid on the project. Later in the certification section we will see how these controls are set up.

## 9.5 Design

In the *design* stage, one looks at taking the information from the site survey and adding the network logical and physical designs to it. The site survey deliverables for a project include a coverage drawing, access point locations, power settings, channel settings, and antenna usage. There are no logical IP addressing, cable paths, or wired network changes in their package, although most site survey vendors will find the nearest communication closet and give an estimate as to the cable length needed.

This is where the design phase comes in; a person or a team will figure out the needed IP addressing and other needed changes for the wired and wireless networks. They will provide drawings on the cable paths and might even provide fiber drawings. They will provide an end-to-end view of the network, tying in the new access points and any other devices that might be needed to support the wireless network.

When looking at the scenario with MoreCo, the best option is to have the local support staff at each site perform the first attempt at the design. After that has been completed, a review will take place with the site survey vendor, project manager, in-house consultant, and the previous project designer. During this review, a number of other people whose eyes looked over the design will be able to bring a new perspective as well as old lessons learned to the table. This will create a smoother flowing design phase that gets faster and more consistent as it progresses. After the review, the designer will update his package and a final round of review will occur, this time with a limited number of attendees.

## 9.6 Staging

Next comes the *staging* portion; this can be on-site or at a central controlled location. The equipment needed for a project must be turned on, configured, and tested. This testing is often easier to do with everything set up in one location than trying to go from one end of a building to the other every time you want to test or make a needed change. Once everything has been correctly tested for hardware failures, it can be tested as a whole (or what is called a system test). This test will verify any redundancy built into the design and cause the network to perform as it should in the event of a failure. If the staging takes place in a central location, the added benefit of having the same group or team see all the network designs in relation to a project helps keep consistency across the board. This step will ensure uniformity and repeatability across the enterprise.

Looking at MoreCo, the choice was to create a project office where they have a team centrally configure all the new equipment. This allowed them to catch small inconsistencies that create large headaches in the field. This approach also allows for a very well-defined build process, which saves time compared to performing this function at each location with different team members.

## 9.7 Deployment and Installation

Once the equipment is staged, it should be installed. Any new network electronics can be easily racked and stacked. Performing the physical design helps prevent a technician from going to the rack to stack the electronics and finding out there is no room in the rack. Installing the access points can also be tricky. They must be hung in close proximity to the site survey locations. Some site survey vendors will leave a small strip of fluorescent tape identifying where the access point should go. The site survey package should include documented designs showing where and how each access point should be hung. To install them, there are a number of options: perform it in-house; use a contractor; or in some cases, have a union member perform the work. The key is to define how to mount the access points and antennas. Many times, this will create problems because the person doing the mounting has no technical education as to antenna orientation. That person also has no idea of what not to mount them next to. This is where antenna-mounting details come into play; these details must be provided by the site survey vendor or created as part of the physical design.

Some installation mistakes are very common. Hanging omni antennas against columns or walls is a grave mistake. Omni antennas should be installed at least three feet away from any column. They should never be installed next to or on a wall. Another big mistake involves antenna orientation. Yagi antennas may need a slight tilt downward; they also need to point in a certain direction to be effective. Patch antennas have similar requirements. They may need to be flipped over; sometimes, they are installed upside down. The reasoning behind this is that they have an 8° tilt toward the antenna leads. Another serious issue surrrounding mounting involves obstructions; when a site survey vendor does its survey, something might not be there, such as a wall or a big metal duct, that is present now. Another mistake that can be prevented is to require a 50' cable loop in order to easily move an access point if there is a coverage issue. Some people say that this loop can cause induction, although no testing has surfaced proving this. If this concerns a team, they can always install this loop at least 20 feet away from the access point, somewhere along the cable run.

In looking at MoreCo, one sees that they chose to have their local site maintenance staff hang the access points. They received all the needed drawings from the site survey vendor and were under the watchful eye of the project manger. This helped ensure that none of the mistakes listed above happened; and with the utilization of the same project manager, lessons learned carried over to the various sites.

## 9.8 Certification

After the system has been fully installed and is online, it must be certified by a third party and the site survey vendor. The third party ensures that the site survey vendor has not overlooked any gaps or possible issues before signing off on it. The third party can be the contracted consultant or even a local site member. This is when the site survey company would end its warranty and lift the financially responsibility for any coverage issues. This is why having a third party perform a certification can help identify any gaps or holes in the coverage area. This is why having a third party perform a certification can help identify any gaps or holes in the coverage area. This is also the time for the site survey vendor to make sure all the access points are installed where wanted, based on their design and documentation. The step is often performed in tandem with many groups that are involved. The site survey vendor, project team, end user, and installation crew are all likely attendees for this step. Not all of these groups or people will need to perform the certification, although if any problems arise, such as incorrect mounting, each team is on-site and ready to make the changes.

Next is performing the actual certification. This is often done by following the survey process again, although this time the access points are installed so most of the equipment is not needed. While this is being performed, the site survey vendor is releasing his requirement to fix any coverage issues as each access point is verified to have adequate coverage per the design. Other than coverage, tests are most often performed using the actual wireless clients. This testing ensures that the entire system works and functions at an adequate level while all players are present.

At MoreCo, they decided to have their independent consultant perform the third party certification with the site survey vendor. This ensured that the site survey vendor did not perform any work that was not up to the expectations of the consultant, as well as the company. This also gives the site survey vendor a chance to show correct functionality and get a sign-off on their installation. This sign-off will state that the required areas do in fact have adequate coverage. The MoreCo team performed testing of their devices to ensure that the system as a whole operated correctly.

## 9.9 Audit

The audit phase is similar to the certification; although its main objective is not to have everyone sign off on a project as complete. The audit objective is to make sure nothing was done outside of any existing corporate standards and that the network is secure according to corporate security policies. The audit process also ensures that after the network is deployed, it is not changed in any non-standard way. In some remote sites people are often removed from corporate headquarters and are not as forthcoming to ensure compliance. Because of this, security is often pushed away in favor of easier administration or support. This is one of the main reasons why an audit is so critical to the enforcement and assurance that networks are maintained in a compliant fashion.

## 9.10 Chapter 9 Review Questions

1. What is the second step in a wireless deployment process?
   a. Business case
   b. Design
   c. Certification
   d. Estimation

2. When performing a survey, what should the surveyor NOT do?
   a. Locate any interfering devices
   b. Correctly place access points
   c. Provide adequate coverage
   d. Use local power

3. When performing a survey, what should a survey team always assume?
   a. More work than expected will be needed
   b. Everything will break
   c. Access into a building will be taken care of
   d. Access point might fail

4. When creating a business case, what piece of data is most important?
   a. Requirements
   b. Area of coverage
   c. Estimated cost
   d. Time frame

5. When creating an RFQ for a site survey, when should the warranty of the site survey end?
   a. Business case
   b. Design
   c. Certification
   d. Estimation

6. What is the first step in a wireless deployment process?
   a. Business case
   b. Design
   c. Certification
   d. Estimation

7. What is the final step in a wireless deployment process?
   a. Business case
   b. Design
   c. Audit
   d. Estimation

8. What item is not normally included in a site survey kit?
   a. Access point
   b. Telescoping pole
   c. Battery to power access point
   d. GPS device

9. What is a spectrum analyzer used for?
   a. To find correct coverage patterns
   b. To locate interference in the 2.4-GHz range
   c. To find the correct channel to set the access at
   d. To locate the next location to install an access point

10. Which item inside the physical information document is not correct?
    a. Location of access point
    b. Placement of antenna
    c. IP Address of access point
    d. Power settings
    e. Channel settings

11. What step of the deployment process is used to validate the site survey vendor's coverage and work?
    a. Business case
    b. Design
    c. Certification
    d. Estimation
    e. System test

# Chapter 10

# Wireless Access Points

This chapter provides an overview of the many different types of wireless equipment. It predominantly focuses on local area wireless equipment: wireless devices, who makes them, what technologies they support, and in what scenarios they should be used. There are many wireless manufacturers out there, so many that an entire book could be put together to look at each of them in detail. This chapter attempts to address as many as possible, going in-depth only on the most commonly manufactured types.

The single most important piece of equipment in a wireless network is the access point. The access point is the piece of equipment that propagates the wireless signal into the air. There are many types of access points from many different vendors. Some of them are made for small office/home office (SOHO) and some are made for large enterprise deployments. Cisco is one of the biggest players in the access point market; others include Proxim, Aruba, Symbol, and SonicWall. As discussed previously, there is an overwhelming number of manufacturers trying to sell access points. Therefore, the focus stays with some of the most used products in the industry.

## 10.1 Linksys Access Points

Linksys was acquired by Cisco in 2004. Thus far, Cisco has decided to keep Linksys separate as a SOHO-only product line. This means that the equipment that Linksys has is primarily targeted at SOHO environments.

One interesting thing that Linksys has done is move most of its wireless products to an open source Linux kernel. A full list of all the Linksys access points with open source firmware is located below. This has allowed Linksys to utilize some already-existing open source code to achieve lower development costs.

Using an open source code has sparked many groups who have hacked into Linksys access points and made some interesting features available within their own firmware code releases. Some of the most common hacked firmware types are Wifibox, Batbox, and Alchemy. These firmware versions are rather easy to install and can quickly improve many of the features of a Linksys access point. Some of them have added functionality that was not present in the Linksys firmware releases.

Access points with open source firmware include:

■ WRT54G
■ WRE54G
■ WPG54G
■ WET54G
■ WAP55AG
■ WAP54G

Linksys currently has more than 12 access point models. Most of the access points look like the one shown in Figure 10.1. Some newer models have started to take on a more compact design like the one shown in Figure 10.2. This section looks at each of the two styles and sees how they function.

Looking at Figure 10.2, one can see the older body style. Linksys has ten access point models that follow this body style. Although this is the older body style, it is more likely that one will run into this access point style before the newer style is out in large numbers. The older Linksys access point has a number of indication lights. Each of these lights allow for an easy glance at what is working and what is not. Figure 10.3 provides an up-close look at these lights.

Lights appearing from left to right in Figure 10.3 include:

■ *Power light.* This light indicator informs the user that power is being received at the access point. This light also signifies that the device is able to start the boot-up process. It will stay on the entire time the unit is powered up to indicate operation. When the unit is undergoing its self-diagnosis testing during boot-up, this light will flash.
■ *Wireless light.* This light indicates that there is a wireless connection. It will blink if wireless traffic is passing through the access point.

**Figure 10.1   Older-style Linksys access point.**

**Figure 10.2   Newer-style Linksys access point.**

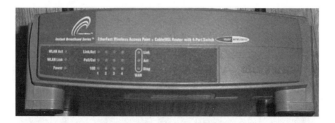

**Figure 10.3** Older-style Linksys access point front LED detail.

**Figure 10.4** Older-style Linksys access point side port detail.

■ *Port 1–4 light.* This light indicates that there is a layer two connection to a wired device on one of the four wired ports of the device. When one of the connected devices starts to communicate, the light will blink.

■ *Internet light.* This light indicates that an Internet connection is made to the WAN port of the access point.

The older-style Linksys access points have many buttons, ports, and connections. Going from left to right in Figure 10.4, one can see all of the items detailing each one. First, on the far left, there is a reset button used to restore the configuration of a device to the factory defaults. Next, there is an Internet port used to attach an Ethernet cable to a high-speed Internet connection such as DSL or cable Internet. This port will allow

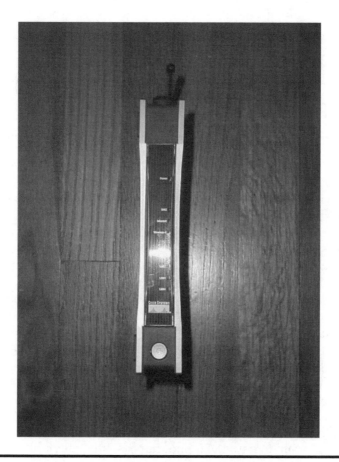

**Figure 10.5   Newer-style Linksys access point front LED detail.**

the use of any Ethernet-capable, high-speed access device. The next items are the four wired ports. This group of 10/100 wired ports allow small offices to connect up to four wired devices to the access point. The final item is the power connector.

Now on to the newer-style Linksys model. The front panel has some of the same lights shown in Figure 10.3, although some other lights have been added to help improve troubleshooting the device. Figure 10.5 shows the lights on the front of the access point.

Lights appearing from left to right in Figure 10.5 include:

■ *Power light.* This light indicator informs the user that the access point is receiving power. This light also signifies that the device is able to start the boot-up process. It will stay on the entire time the unit is powered up to indicate operation. When the unit is undergoing its self-diagnosis testing during boot-up, this light will flash.

**Figure 10.6   Newer-style Linksys access point side port detail.**

■ *DMZ light.* This light indicates the use of a DMZ device. If no device is using the DMZ capability, it will blink during its self-diagnosis testing, as the unit boots up; once completely up, the unit will stop lighting the DMZ light, unless a device is set up to use the DMZ.

■ *Internet light.* This light indicates that an Internet connection is made to the WAN port of the access point.

■ *Wireless G light.* This light indicates that there is a wireless connection. It will blink if wireless traffic is passing through the access point.

■ *Port 1–4 lights.* This light indicates that there is a layer two connection to a wired device on one of the four wired ports of the device. When one of the connected devices starts to communicate, the light will blink.

The back of the newer-style Linksys access point is shown in Figure 10.6. Going from left to right one can see that there is an Internet connection used to access other networks. Next is the four-wired network ports used to connect other wired devices to the network. The first button on the back is the reset button, which is used to reset the configuration to the factory default. The final item is the power connector.

Linksys access points use a Web interface to configure them. To access it, a user must type the IP address of the access point into an Internet browser such as Internet Explorer. When the user does this, he is prompted to enter a username and password. On Linksys access points, the default username is admin and the password is admin. Once the user has logged in to the device, a set-up screen appears like the one in Figure 10.7. This will allow the user to set up the access point. This screen allows a user to set up the WAN connection identifying the network information needed to access other networks. It will also allow a user to set up the LAN network information. This device can be a DHCP server, allowing dynamic network configuration information to be pushed out to connecting devices. This screen also allows for a number of advanced features, such as port forwarding, firewall filtering, and MAC-based security.

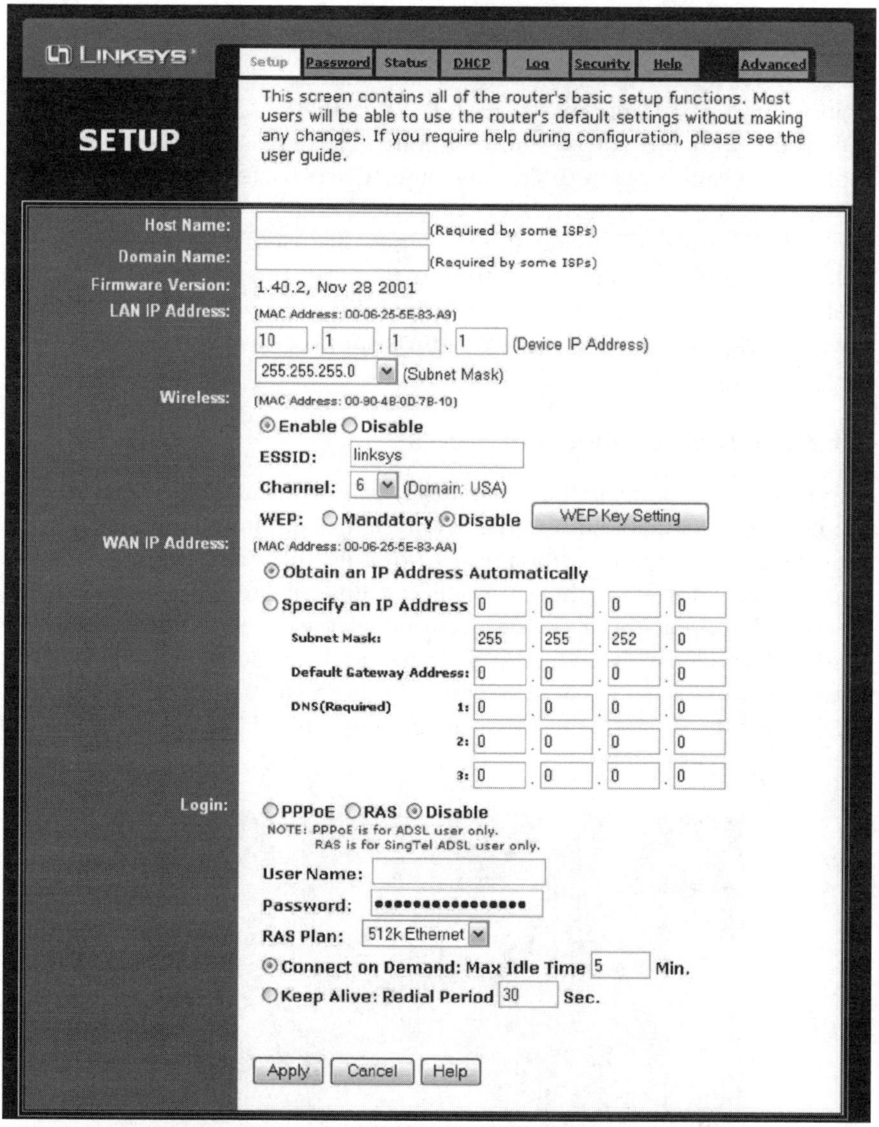

**Figure 10.7   Linksys management screen.**

# 10.2  Cisco Access Points

Cisco has had a stronghold on the enterprise wireless market since its acquisition of Aironet in 1999. Cisco has created multiple product offerings that fit a large array of needs. Once Aironet and Cisco became the same company, their product line began to move toward Cisco's own code.

Once this happened, the market began to see the major benefit that Cisco's wireless access points bring to the table. They allow anyone familiar with Cisco router and switch configurations to understand access point configuration with ease. This was due to the fact that Cisco had integrated its IOS code with the Aironet access point, creating an access point that looked, felt, and operated like any other Cisco router or switch.

Cisco has three main models of access points: 350, 1200, and 1100 series. Each comes in a wide array of types that use different technologies. For example, the 1200 can use 802.11b, 802.11g, and 802.11a. This section looks at all three types of access points and details what connections they have, what protocols and standards they support, and how they operate.

### 10.2.1 Cisco Aironet 350 Series

The first access point is the older Cisco 350 series lineup. These devices are no longer available through Cisco Systems, although they are still very prevalent across many companies. Currently, the Cisco 350 devices are end of sale. However, they are still not end of life, which means that Cisco will support them. When these devices were sold, they came in a hardened model with a metal casing and in a plastic model that had the antennas affixed to the access point. Both 350 models only work with the 802.11b technology, and that technology only. Figure 10.8 shows the 350 series access point.

**Figure 10.8   Cisco 350 model.**

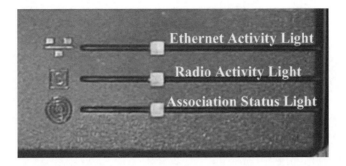

**Figure 10.9   Cisco 350 front LED detail.**

Both of the 350 series access points have the same light and connector layout. The only difference between them is the hardened shell and the ability to affix external antennas to the hardened model. To note is that the plastic 350 series access points were available with the option of affixing external antennas, although this was quickly replaced by the hardened cases. Looking at the front of the 350 series access point, one can see in Figure 10.9 that there are three lights. Each of the functions of the lights is detailed below.

The lights going from top to bottom in Figure 10.9 are:

■ *Ethernet activity light.* This light indicates the status of the Ethernet network connection. The indicator will blink green when a packet is received or transmitted over the Ethernet infrastructure. It will blink red when there is no connection to the Ethernet network.

■ *Association status light.* The association status indicator light signals the operational status of the device. When the light is blinking green, it indicates that the access point is operating normally, although it is not connected to any wireless clients. When the light is steady green, it indicates that the access point is associated with at least one wireless client.

■ *Radio activity light.* The radio indicator blinks green to indicate the presence of radio traffic activity. It is usually off unless there is traffic on the wireless, at which point it will blink green.

On the back of the 350 series access point is an RS-232 connection for terminal emulation and a single Ethernet port. On the hardened access point, there are two RP-TNC connectors that attach to external antennas. Figure 10.10 details the connections located on the back of the Cisco 350 series wireless access points.

**Figure 10.10   Cisco 350 side port detail.**

**Figure 10.11   Cisco power injector detail.**

The 350 series access point has no power plug for the unit. This is because the unit works off Power over Ethernet (POE). POE is a method by which power can travel down a network cable. This power is used to power a small device such as an access point or VoIP phone. To use the POE, one must have a switch that can provide POE. If no such switch is available that can support POE, one can use the small power converter that ships with the access point. This small power cord takes a Category 5 connection from the network and another Category 5 connection from the access point and provides power down the cable. One note about using power injectors is physical real estate. In larger companies, physical rack space is at a premium. Using a power injector means that a physical device like the one in Figure 10.11 needs to exist between the telecom-munications closet and the access point. The most logical place to put this device would be inside this closet on a shelf. The physical real estate needed for this shelf and power injector is one of the most commonly missed items when planning for a Cisco wireless network.

Looking at Figure 10.11, one can see the power injector that comes with the Cisco 350 access point and one that comes with the 1200 series access point. They have a power connector that is used to connect external power to the injector and two Ethernet connectors (one for a connection to the network, which is not powered, and one connection to the access point, which is powered). The DC voltage used to power the access point is −48 VDC. There is a newly adopted POE standard called 802.3af. Some of the older access points are not considered 802.3af compliant. Cisco states that these devices will work correctly on its POE solutions, although to use another vendor's 802.3af-compliant switch the access points must be 802.3af compliant. Today, all the access points shipping from Cisco are 802.3af compliant.

The operating systems on these units are unlike the familiar IOS that Cisco bases most of its products on. This was due to the acquisition of Aironet; along with acquiring the company, Cisco also acquired the access point code. This meant that the setup and management of the access point is different from most other Cisco devices. To fix that, Cisco needed to change the code into IOS. In late 2003, Cisco created a software code release that converted the old VxWorks operating system into IOS. Thus, in this subsection, the focus is on VxWorks for the 350 series only. This is because when we talk about the 1200 and 1100 series, it is obvious that they only use IOS (IOS is discussed later).

Connecting to the access point can be accomplished in a number of ways. First, to access it without any knowledge of its setup, one needs to access it though the console. This is the RS-232 connection on the back of the device shown in Figure 10.10. A serial DB-9 cable comes with the access point. It is blue and should have two ends that look alike. To note is that most Cisco equipment comes with what is called a rollover cable. This cable is used to access the equipment through the console port. The 350 series access points are one of the few Cisco devices that have a different cable for console management than the normal Cisco console cable. Once this cable is connected to both the access point and an open Com port on a workstation, a terminal emulator program must be open. One emulator that is part of Microsoft Windows is called hyper terminal. Other emulators can be downloaded from the internet or purchased. Make sure the terminal emulator program is correctly configured following the details below.

To connect a workstation correctly to an access point, the setup on a terminal emulator program needs to have the following settings:

- *Com Port:* set this to 1 or the com port to which the cable is connected
- Bits per second (baud rate): 9600
- Data bits: 8 bits
- Parity: no parity
- Stop bits: 1 bit
- Flow control: Xon/Xoff

```
AP350-57f029      [Cisco 350 Series AP 11.23T]      Uptime: 06:27:44
------------------------------------------------
Associations
   [Clnts: 0] of 1   [Rptrs: 0] of 0   [Brdgs: 0] of 0   [APs]: 1
------------------------------------------------
Events
    Time        Severity         Description
------------------------------------------------
Network Ports                               ===[Diagnostics]===
   Device              Status      Mb/s      IP Addr.      MAC Addr.
[Ethernet]             Up         100.0    10.0.0.50     00409657f029
[AP Radio]             Up          11.0    10.0.0.50     00409657f029
------------------------------------------------
Home - [Network] - [Associations] - [Setup] - [Logs] - [Help]
[END]


(Auto Apply On) ^R, =, <ENTER>, or [Link Text]: s



AP350-57f029                    Setup                  Uptime: 06:27:46

===[Express Setup]===
                             Associations
[Defaults Associations]      [Address Filters]              [Advanced]
                             [Port Assignments]
[Ethertype Filters]          [IP Protocol Filters]          [IP Port Filters]

                             Event Log
[Defaults Event]             [Event Handling]               [Notifications]

                             Services
[Console/Telnet]  [Boot Server]      [Routing]          [Name Server]
[Time Server]     [FTP]              [Web Server]        [SNMP]
      [Cisco Services]        [Security]        [Accounting]

                          Network Ports          ===[Diagnostics]===
[Id Ethernet]     [Hw Ethernet]      [Fltr Ethernet]    [Adv Ethernet]
[Id AP Radio]     [Hw AP Radio]      [Fltr AP Radio]    [Adv AP Radio]
------------------------------------------------
[Home] - [Network] - [Associations] - Setup - [Logs] - [Help]
[END]
(Auto Apply On) ^R, =, <ENTER>, or [Link Text]: █
```

**Figure 10.12   Cisco 350 VxWorks console screen.**

Once the cable is connected and the terminal emulator program is set up correctly, try to launch the session and connect the access point. At this point, one should see a cryptic-looking menu system similar to that in Figure 10.12. This is the VxWorks operating system. One can navigate by typing the first characters of a command into the window. The commands have brackets around them; some of them have different navigation keys than their names imply. Look out for what is typed in the brackets on the screen to make sure that the correct information is entered.

Having learned how to connect to the Cisco 350 access point via the console, there some other ways of setting up and managing this device. This device has a built-in Web interface like the one on the Linksys access points. To access it, one must know the IP address of the unit so one can enter that into a Web browser. By default, all Cisco access points get their address via DHCP. This makes finding the access point's IP address difficult the first time. Cisco stepped in and created a tool called IP setup utility (IPSU) that can find the access point's IP address from its MAC address. For this tool to work, it must be installed on a workstation that is on the same network segment as the access point. Once this software is installed, one can launch the application, type in the MAC address of the access point, and it will show the IP address. The MAC address of any Cisco access point is written on the back of the unit.

Now that the IP address is known, connecting to the access point is possible through the Web. To do so, just type in the http:// and the IP address of the access point. This will bring up the Web interface. For VxWorks, this method is the preferred method of configuration. One key advantage that Cisco has made for itself in the large enterprise space is the ability to create configuration scripts for almost all of its products. With VxWorks, still being an Aironet/Cisco code, this was not achieved. This meant that the ability to easily script the configuration in VxWorks was rather difficult. Looking at Figure 10.13, one can see what the Web interface looks like for the VxWorks operating system.

The final method of accessing the access point's management functions is through telnet. This has a very similar look and feel to the console although it can be done remotely from any connection with IP connectivity. In some newer versions of code, Secure Shell (SSH) can also be used. Telnet is prone to eavesdropping because its authentication takes place in cleartext. To telnet into an access point, all one needs is a telnet program. UNIX and Windows both have telnet ability right from a command line or shell. Most Cisco equipment can perform a telnet action from one device to another.

## 10.2.2 Cisco 1200 Series Access Point

Now to the Cisco 1200 series access points. These all come in a plenum-rated metal case; the case is shown in Figure 10.14. These access points have the latest and greatest features available from Cisco. They are capable of supporting 802.11b, 802.11g, and 802.11a simultaneously. To support 802.11a, a paddle card must be installed into the access point. This card is shown in Figure 10.15.

Delving deeper into the access point itself, one can see from Figure 10.14 that the access point has the same LED layout as the 350. Each

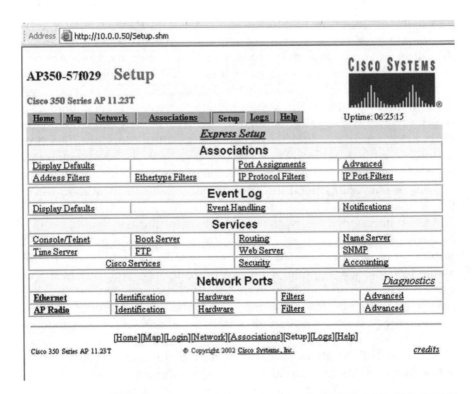

**Figure 10.13    Cisco 350 VxWorks Web access screen.**

Ethernet activity, association status, and radioactivity light performs the same function as detailed above for the 350 series. Looking at the connections on the device in Figure 10.16, one can see that the 1200 access point has two antenna leads, a power connection, an RJ-45 console port, and an RJ-45 Ethernet access port. One big difference between the 350 and 1200 series is the more common use of the correct Cisco-style console cable. The 1200 access point has the same Cisco console cable that almost all other Cisco equipment has. Another big difference is the directly connected power port. This is now included in this model, unlike the 350, which required POE from a switch or injector. The 1200 series access point is capable of POE and can be plugged in with direct power from a power cord. This allows for maximum flexibility with regard to powering the device. Also, all currently shipping 1200 units from Cisco are 802.3af compliant.

The 1200 series access point was made with the capability to upgrade to new technologies. This was often a requirement of customers. They wanted the capability to support technologies that were close to being released, although not available at the current time. To make sure that

**Figure 10.14    Cisco 1200 access point detail.**

they did not have to replace all their wireless networks, most customers required that the access point have easy upgradeability, to include support for other, newer technology. In the case of the Cisco 1200 series access point, this change was as easy as swapping out one card for another. This allowed the device to go from 802.11b to 802.11g rather easily. When the 1200 first came out, 802.11g was still being standardized. This meant that Cisco had to release the 1200 with 802.11b technology. To upgrade to 802.11g, a new radio must be installed. Cisco wanted to make the transition from 802.11b to 802.11g an easy one and did so by making it easy to change out the radio cards. Looking at Figure 10.17, one can see how the radio comes out.

The 1200 series access point runs IOS. Most Cisco products run this operating system. It is laid out the same way as other Cisco devices. This means that anyone familiar with a Cisco router or IOS-based switch should be able to pick up the access point IOS without too many issues. This IOS still has the ability to allow connections from the console, telnet, Web browser, and SSH. One of the big differences in IOS versus VxWorks is the ability to script configurations. With IOS, text configurations are easily

**Figure 10.15   Cisco 1200 removable 802.11a radio detail.**

**Figure 10.16   Cisco 1200 side port detail.**

put into and pull out of any 1200 IOS-based access point. This means that basic standards can be inserted into engineering templates and some assurance can be given that all access points will share these common settings.

Figure 10.18 reveals the IOS Web-based interface. It looks very different from the VxWorks. It has all the same functions of VxWorks, although it is

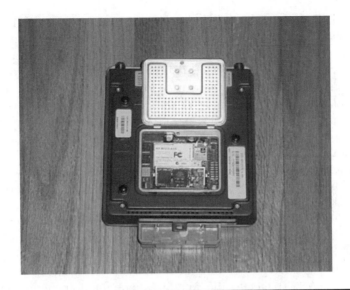

Figure 10.17　Cisco 1200 radio replacement.

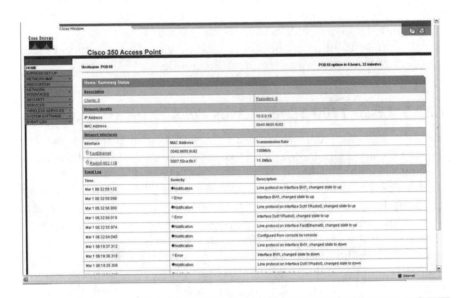

Figure 10.18　Cisco 1200 IOS-based web access screen.

laid out a little neater. When using this configuration tool, one of the quickest ways to get the access point up and running is the express setup page.

When one consoles or telnets into an access point, one gets what is called a command line interface (CLI). This is very similar to a Cisco router. Once logged in, one is in user mode, which is a very restrictive mode and allows for only simple viewing commands. To get into the next privilege mode, called the EXEC mode, one must type "enable" into the command line. This should change the prompt from > to #. Now this mode can run all viewing commands. If one wants to make a change, one must enter another mode, called global configuration mode. This mode allows one to make changes. To get into global configuration mode, one types "configuration terminal" into the command prompt. One can make changes that apply to the entire device, including changing the device's name, setting the clock, adding users, etc. The final menu type is called interface configuration mode. This menu allows one to perform actions on certain types of interfaces, such as the 802.11b radio, or a particular VLAN. To get to the interface configuration menu, one must already be in global configuration mode and then type "configure interface dot11radio0." This gets you into the configuration menu for the 802.11b radio. Whatever designator is assigned to the interface one wants to configure is the one that must be referenced in the command. Figure 10.19 provides all the details and shows what the IOS command line looks like. Looking at this figure, one can see that one of Cisco's features is support for the tab command. In most UNIX operating systems and Cisco IOS, the tab key can be used to complete a command. As one can see from Figure 10.19, to enter enable mode, all that is needed is "en." This can also apply to other menu items such as global configuration mode; typing in "config t" can easily access this mode.

### *10.2.3 Cisco 1100 Series Access Point*

The final wireless access point from Cisco discussed herein is the 1100 series. This access point was created for a low-cost, small office access point. It is only available in 802.11g or 802.11a. Unlike the 1200 series, it cannot run two radios at once. Older 1100 access points ran 802.11b standard. To upgrade from 802.11b to 802.11g, one only needs a small radio chip. This makes the 1100 a modular upgradeable radio. Both the 1200 and 1100 series access points only need a single screw to get to each of their radio cards. The 1100 series access points do not support external antennas, unlike the 1200 series, which does support external antennas. The 1100 has a 2.2-dBi omni antenna permanently affixed to the unit. This makes the unit difficult to use in places where external

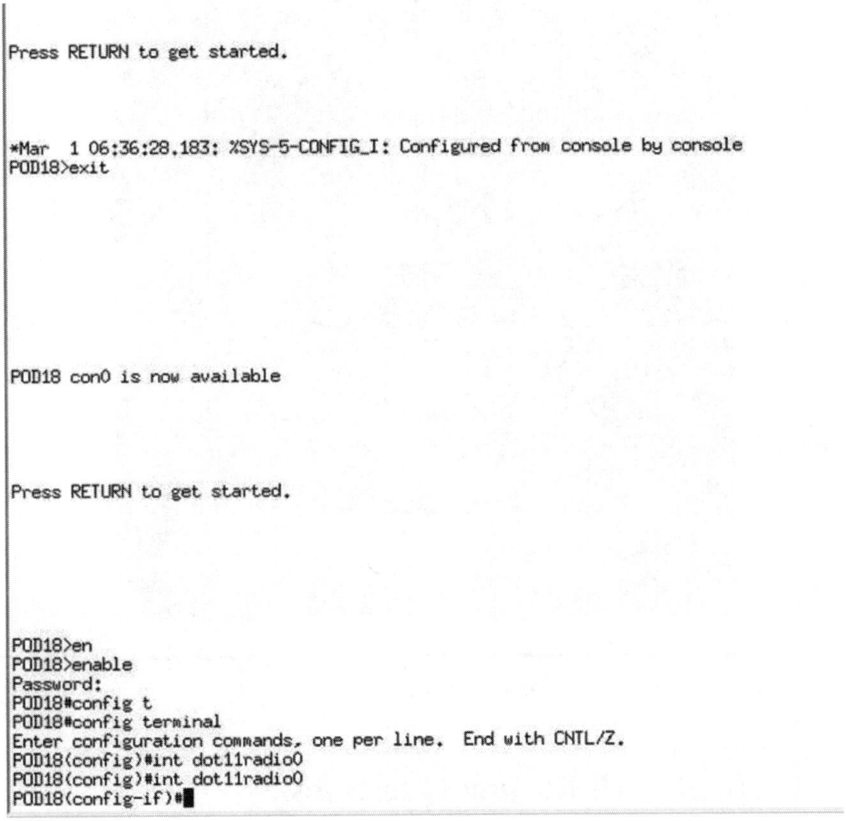

```
Press RETURN to get started.


*Mar  1 06:36:28.183: %SYS-5-CONFIG_I: Configured from console by console
POD18>exit




POD18 con0 is now available




Press RETURN to get started.




POD18>en
POD18>enable
Password:
POD18#config t
POD18#config terminal
Enter configuration commands, one per line.  End with CNTL/Z.
POD18(config)#int dot11radio0
POD18(config)#int dot11radio0
POD18(config-if)#
```

**Figure 10.19   Cisco 1200 IOS console screen.**

antennas would be needed. If one needs an access point that requires external antennas, go with the 1200 series. If the solution is for a small office, then the 1100 series is a more cost-effective option. The 1100 series access point is capable of being powered by POE. Looking at Figure 10.20, one can see that the 1100 series access point looks similar to the 1200, although it is slightly smaller and has affixed antennas.

Other than the many physical differences between the 1100 and 1200 series access points, from software perspective the two are very similar. Both run IOS and operate under the general command structure. This makes an 1100 and a 1200 look almost the same from a remote console perspective. This also helps with large companies that manage both 1200 and 1100 series access points. They both look and feel the same from a software configuration standpoint.

**Figure 10.20 Cisco 1100 access point detail.**

## 10.3 Chapter 10 Review Questions

1. What is the correct DC voltage used to power a Cisco 1200 series access point when using POE?
   a. −23
   b. −34
   c. −99
   d. −48

2. The cable used to connect to a Cisco 1200 series access point is commonly referred to as a _____.
   a. Patch cable
   b. Coaxial
   c. Roll down
   d. Roll over

3.  What administrative methods are supported in a Cisco access point?
    a.  IPSSU
    b.  Console
    c.  Telnet
    d.  Web based

4.  Which of the following is not a connection on the Cisco 1200 series access point?
    a.  RJ-45
    b.  Power
    c.  Fiber
    d.  Ethernet

5.  Which types of antennas does the Cisco 1100 series access point support?
    a.  13.5 dBi Yagi
    b.  5.5 dBi patch
    c.  2.2 dBi omni
    d.  6 dBi patch

6.  What administrative methods are supported in a Linksys access point without firmware modifications?
    a.  IPSU
    b.  Console
    c.  Telnet
    d.  Web based

7.  What command should be issued on a Cisco IOS access point after accessing it from a telnet or console to reach the enable mode?
    a.  Int Dot11radio
    b.  Config t
    c.  Enable
    d.  Configuration mode

8.  What types of operating systems do new Cisco access points use?
    a.  IOS
    b.  VxWorks
    c.  Microsoft
    d.  Linux

9. Do Linksys access points support POE?
   a. Yes
   b. No

10. Does Cisco currently sell 350 series access points?
    a. Yes
    b. No

11. Cisco 1200 series access points are capable of supporting 802.11b, 802.11g, and 802.11a all at the same time.
    a. True
    b. False

12. What standard do all Cisco 350 series access points adhere to?
    a. 802.11a
    b. 802.11b
    c. 802.11g
    d. 802.11n

13. What Cisco access point runs the VxWorks operating system?
    a. 1200
    b. 1100
    c. 350
    d. 1400

# Chapter 11

# Wireless End Devices

When looking at wireless end devices, one notes that laptops are far from being the most common. Depending on how wide the definition of wireless end devices is, cell phones could easily top the list. Back in the pre-802.11 days, most wireless devices were handheld scanners used for inventory tracking. As technology advanced and data rates increased, a new market for wireless laptop communication evolved, thus adding wireless laptops to the list. With the continued advancements in wireless technology, more and more devices began to incorporate wireless connectivity. During this time, PDA devices began to hit the market with wireless capabilities. Today we have multiple communication capable devices. These smart phones and PDA devices have Wi-Fi, cellular, Bluetooth, and IR technologies that can all be used to transmit data to and from a single device.

This chapter looks at the most common wireless end devices, explaining how they work, what purpose they serve, and how security relates to them. After exploring all the wireless devices, this chapter then explores what threats face each of these devices and how one can protect them from those threats.

Before getting into the specifics of each type of wireless device and their related risks, it is necessary to explore the risks of wireless devices in general. Almost all wireless devices are mobile; unfortunately, this means there is a risk of lost or stolen devices. Mobile devices such as laptops, PDA devices, and others are often misplaced or, worse, stolen. This can mean the device itself is stolen or the data inside it is stolen. When this happens, the loss of an asset has occurred. Depending on the price of the device, and the data that may have been on it, losing these devices can be a real risk to any organization. To protect against this risk, companies can use encryption software for the protection of their data and good policy development to combat device loss. The most important risk-reducing action is education and awareness.

## 11.1 Laptops

Laptops are the devices people most commonly picture when they think of wireless LAN end devices. Laptops are able to perform all the same functions as a desktop computer. Many people prefer laptops to desktop computers because they are small, portable, and convenient. Once the wireless industry took off, the laptop industry started to see the benefits of going wireless. More and more consumers also began to see the benefits; they could browse the Internet from their couches or anywhere in their homes without the inconvenience of wires. Today, the majority of laptops come with a built-in wireless adapter. Laptop computers connected to wireless networks face not only the security threats all network connected computers face, but also the added threats from wireless access.

One example of a threat that wireless laptops face is a hotspot. Using a laptop on a hotspot might seem safe, but the truth is, it is far from safe. The way most computers communicate on a network is by network protocols; some of these protocols were made back in the early days of the Internet itself. The age of these protocols can lead to a host of security issues. This is magnified when using a hotspot, especially one that does not use encryption. These older protocols are more exploitable in a wireless environment than in the physical media of a wire.

Picture the following example of how wireless laptops are at risk. A person goes to his local hotspot. It could be in an airport, coffee shop, bookstore, or practically anywhere. The laptop user fires up his laptop and connects to the network. The user fills out the security information necessary to gain access and the hotspot proxies all his requests to the Internet. Because the hotspot is doing the security at the access point, the airwaves are not secure in any fashion. If a hotspot vendor was to secure the airwaves with some form of encryption, it would be more of a risk to the vendor than an advantage for its customers. If the hotspot vendor says his network is secure and an incident happens, the vendor could be at fault and liable for all damages. Because of this liability, most hotspot vendors protect only Internet access and leave the airwaves open. Now that the laptop user has connected to the Internet, he will most likely want to retrieve his e-mail. The most common e-mail retrieval protocol is POP3 so one assumes that is what he uses. The POP3 protocol has quite a few security deficiencies; for example, it sends the username and password in cleartext. This means that anyone in the area using a network sniffer can capture the username and password of someone else downloading mail. Once the name and password have been captured, the attacker is able to log into the laptop user's mail system and read, write, or delete his e-mail. Other protocols have problems similar to POP3. This book explores a number of weak protocols.

## 11.2 Tablets

Tablets have slowly begun pushing their way into the marketplace. Many industries were unable to function with the small form factor of PDAs but did not like the weight or two-handed use of laptops. Tablets were created to mimic a notebook with the advantage of having all documentation and handwriting electronic. For example, this has helped healthcare become more streamlined. Tablets have moved forms and documents that need signatures from paper base to electronic factor. The tablet is nothing more than a computer with a compact form and an electronic pen. This means it is also susceptible to all the threats mentioned in the above sections.

An example of a threat related directly to tablets involves the use of electronic signatures. When a tablet is used to electronically sign a document, the signature becomes software based. This means the signature can be cut, copied, and pasted into a different form. This can mean that someone can easily forge signatures. This threat is not directly related to wireless use; however, it is very common on tablet computers.

## 11.3 PDA Devices

Information management tops the list of features for PDA devices. For the majority of the 1990s, PDA devices were marketed as only information managers. As the turn of century approached, these devices began to incorporate new technology that pushed their features way beyond that of a personal information manager. With the advent of wireless technologies, PDA devices have become useful for multiple business applications. Some of these involve database querying, record keeping, record viewing, and other real-time tools. These new abilities make PDA devices practical and cost effective for many enterprises.

All PDA devices run on some type of hardware; this hardware enables the device to turn on and use various wireless technologies. Today, there are so many PDA device manufacturers that talking about each of them would easily fill this book. Having all these manufacturers creates both positive and negative effects. On the positive side, having a large selection is always good for consumers. This helps customers find better prices and allows for a wide variety of features. Allowing the consumer to choose the type, speed, form factor, and available transmission technology allows for a quicker purchase.

However, supporting different devices can cost a lot more money than supporting a single one. This has become evident in corporate environments where personal PDA devices are accepted. This culture has brought with it requests from users to have IT personnel to support these devices. Unfortunately, when a company tries to save money by not embracing a corporate PDA roll-out and instead provide end users with the ability to use their own personal PDA devices, the support cost alone will outweigh the hardware costs of an enterprisewide roll-out. With multiple vendors running multiple versions of software, being able to support this gamut of devices is overwhelming.

Even with all the different hardware options out there today, across all the different vendors, none of these devices would work without software. This software allows the device to power on and allows it to interact with the user. Similar to the hardware market, the software market offers a wide variety of options.

### 11.3.1 Palm

The company that makes Palm software was created in 1992. One of their first devices was called a Zoomer. The Palm Company has changed hands several times since its creation in 1992. It was acquired by U.S. Robotics in 1995, then by 3Com in 1997 as they acquired Palm's newly appointed parent company U.S. Robotics. Then, three years later in 2000, 3Com turned Palm back into an independent company, bringing it back full circle to where it began in 1992. In 2002, Palm spilt its operations into two companies owned by the parent company Palm: (1) Palm Source, which created and licensed the Palm OS software, and (2) Palm Solutions Group, which handled the creation of hardware for the Palm operating system.

Today, Palm represents a large portion of the handheld market. However, its market share began falling due to the recent advancements of other companies. Palm devices continue to be primarily personal information management (PIM) devices carrying the user's contacts, calendar, and other useful tools.

The base Palm software comes with the following applications:

- Address book
- Mail

- Date book
- To-do list
- Hotsync software

Some other applications available with newer models include:

- Note Pad
- Short Message Service (SMS)

## 11.3.2 *Microsoft CE and Pocket PC*

Another software maker for handheld devices is Microsoft. Microsoft made a name for itself with the creation of the computer software operating system called Windows. In 1996, Microsoft started introducing its successful desktop software into the handheld computing market. The first version of software released by Microsoft was CE 1.0. This software was similar to Windows 95 and had the same look and feel. CE 1.0 was a slimmed-down version of Windows 95 that was made solely for handheld devices.

As time went on, newer versions of CE began to emerge from Microsoft and new initiatives emerged as well. One of those initiatives was an OS targeted at wireless handheld computers and mobile phones. This initiative created Pocket PC. This left Microsoft with CE and a choice: either keep it going or replace it with Pocket PC. Microsoft took its version track for CE and added it into part of their embedded operating system group. This meant that for devices like bar code scanners, radios, TVs, and other similar devices, the Microsoft CE platform will be an available standardized solution. This split between CE and Pocket PC allowed the Pocket PC group to build more functionality into its smart phone operating systems and PDA devices. Pocket PC was geared more toward a user who needed to perform multiple functions like e-mail, contact management, calendaring, and many other PIM type applications. On the CE platform, the normal user would only be using the device for a particular function, such as bar code scanning or package delivery.

Microsoft allows hardware makers to customize their CE software to operate on their devices. This has helped companies create devices and not software. Many devices have hardware built by a certain company and the software provided by Microsoft. One example is a portable handheld scanner that runs CE. A hardware company makes the scanner and Microsoft provides the software.

Looking at CE today, which is now CE.NET, when compared to Pocket PC, the main difference lies in the PIM software. CE.NET has no PIM software included; there is no calendar, e-mail client, or task manager.

This is because of the change Microsoft made from handheld devices to devices in general. The Pocket PC version that Microsoft released is made just for PDAs and smart phones. Microsoft has talked about splitting smart phone software away from Pocket PC software. The real issue that has stopped this event is the massive push to condense these functions into a single information management and communication device. Currently, if one purchases a PDA unit, it should come with Pocket PC and not Windows CE; if it does not, be careful as this means that the manufacturer is way behind the curve. Some of the applications that come standard on the Pocket PC are listed below. Make a note that the smart phone OS has only a limited subsection of the applications listed below.

Standard applications that ship with Pocket PC include:

- Pocket Outlook:
  - Contacts
  - Calendar
  - Tasks
  - Inbox
  - Notes
- Microsoft Pocket Word
- Microsoft Pocket Excel
- Microsoft Reader
- MSN Messenger
- Terminal Services

## 11.3.3 BlackBerry RIM OS

BlackBerry's RIM operating system encompasses software, hardware, and the needed application to connect the mobile device to corporate e-mail servers sold, all from a single company. The device OS was created from the ground up to have an always-on connection to the Internet and e-mail. Because of this, a solution had to be created to support this type of connection to a corporate e-mail server. Most companies had pieces and parts to this whole solution. BlackBerry created an entire solution based on its own products, which was never done by a single company before then. This solution is based on all of their own hardware and software. There are some major differences between the BlackBerry devices and other PDA versions. First, they have no stylus to navigate; they use a roller on the side of the device to facilitate navigation. They also all come with a small keyboard that allows for easy messaging. Some PDA hardware makers have seen these differences and have adapted some of them to their offerings, such as the keyboard for Pocket PC devices.

## *11.3.4 Symbian OS*

The Symbian OS was created to meet the needs of users who were complaining about how the industry failed to respond to the need for smart phones that can combine PDA, messaging, and voice calling functions on a single device. Most manufacturers or software makers took whatever product they already had and adapted it to fulfill such needs. This meant PDA devices and software tried to tie messaging and voice capabilities into one device. On the other side, phone makers tried to create more PIM-type applications on their devices and in their software. Then a couple of these companies got together and led a joint effort to create software that was originally created to operate on smart message- and voice-aware devices. This led to the creation of the Symbian OS.

## *11.3.5 Linux*

Linux, which provides free-of-charge, community-based operating systems, has only found its way to a few devices. Many advanced users have found ways to port Linux onto their devices, although few manufacturers support or sell a Linux PDA. These devices provide an alternative to licensing software from other companies. The Linux PDA has still a long road ahead before one sees them out in any significant numbers on handheld devices.

## 11.4 Handheld Scanners

In this section we are going to talk about handheld scanners and barcode readers. These devices are from pre-802.11 days when the primary use of wireless was to perform inventory tracking or barcode reading. Some of these devices run on 900 MHz and are way outside 802.11. If you have ever been to a large retail store, you have probably seen the handheld scanner guns the cashiers use to scan items. Most modern-day scanners use wireless networks to operate and function in the rule set of 802.11b. The handheld scanner is also used widely in warehouses or distribution centers.

One big issue holding back these kinds of devices is battery life; several vendors have decided to wait or to not pursue 802.11g or 802.11a because of the additional battery capacity needed to maintain their current runtimes. This means that many of these devices will not move to 802.11g or 802.11a. Another major issue with these devices is the chip set and the ability to handle a more processor-intense security method such as 802.11i. Performing Advance Encryption Standard (AES) on these devices will either not be possible or will dramatically shorten the battery life.

The threats to these devices specifically deal with the lack of advanced security capabilities. Some of these products are stuck with only supporting WEP, which can mean a major risk in itself. Another interesting note in this regard deals with 802.11 FH systems. Some have stated that these systems have a lower risk associated with them, stemming from the fact that one cannot go out to a local computer or electronics store and buy an FH card. This means fewer people are likely to attempt to circumvent any secure mechanisms already in place. This topic must be addressed when weighing risks versus controls of older systems such as the ones that support old technologies.

## 11.5 Smart Phones

Today, smart phones are becoming everyday devices that can serve as a planner, phone, Web browser, and small computer — an all-in-one device. Depending on what the manufacturers call the device, it can be a smart phone or a PDA phone. The industry has classified these devices based on their main function, with smart phones being a phone first and PDA second, and PDA phones being a PDA first and a phone second. This section addresses these devices as smart phones although the industry has made a clear delineation between hardware and software types. Microsoft, for example, has two main flavors of operating systems: Smart Phone OS and Pocket PC OS.

Both operating systems perform the same functions, with the only differences being that the Smart Phone OS was developed specifically for a phone and the Pocket PC OS was originally developed for handheld PDA devices and later adapted by each hardware manufacturer to incorporate phone functions.

These devices bring a slew of complexity and risk into account when using them. The first threat directly related to smart phones is the multiple connectivity options. One of these devices can use Wi-Fi, Bluetooth, and a cellular technology on the same device. When security threats come out and are directed at one of these technologies, the user must protect against each of these threats independently. This means that the user must have a higher level of knowledge and understanding about how each of these connection technologies works and is secured.

The virus threat has plagued computers since November 3, 1983, when Fred Cohen conceived the first computer virus. Today, over 20 years later, there are more than 50,000 viruses crawling the Internet. Just recently, we have started to see viruses directed at mobile devices such as smart phones or PDA phones. The virus threat was always known on these devices, although it was only related to having a mobile device carry a computer virus and then, in the process of synchronizing the mobile device to the computer, it would infect the computer with the virus. With viruses and smart phones, not only are there the issues of removing malicious code, but there is also the threat of having malicious code place calls. A virus can attack a device, infect it, and make that device place phone calls or make requests to the Internet that can cost the user money.

Some of these devices have means built directly into them to combat viruses. The Microsoft Smartphone and Pocket PC OS have the default ability to prompt the user before running any executable code. While this may protect against viruses that are sent directly to the phone, most mobile virus writers are well aware of this and have changed their method of attack. The real threat has now moved from viruses more toward Trojan horses. The mobile virus writers want to get people to accept the executable prompt, inserting the virus into a game or an application the user might want to use does this. When they press OK to the prompt, they allow the Trojan to execute and deliver its malicious payload.

Another amazing feature of some smart phones is GPS. GPS is used for a couple of things on a cellular phone or smart phone. First, a method of locating a cellular phone is needed for emergency services. This is one requirement that has put GPS on most cellular phones. GPS on cellular phones most likely is not true GPS, because true GPS does not work correctly indoors due to a line-of-sight requirement between GPS satellites and the GPS receiver. Some cellular carriers use a technology called gpsOne, which works by having the phone be capable of receiving both a GPS constellation signal and a wireless network signal. The wireless network signal is also broadcasting GPS information to the device. Once the device picks up these signals, the measurements are combined by the location server to produce an accurate position fix. Either way GPS is used on a phone, the risks remain the same.

The first risk that comes to mind is tracking. Some information on GPS tracking was defined in the GPS section of the wide area technology chapter (Chapter 7, Section 7.2). In that chapter, one learned that the GPS system itself could not be tracked. The only way to track a GPS receiver is for another technology such as cellular to transmit the location information. Most phones have not only GPS for emergency services, but also location-based services (LBS). Location-based services offer applications the ability to use real-time location data; this can be used to track a delivery driver or find the closest taxi driver to someone who just called in. As amazing as LBS sounds, the malicious implications as well as privacy concerns can be major issues. Luckily, some phones have the ability to disable this feature.

There have been a number of what-if scenarios discussed about the future of LBS on cell phones. The first one is as follows. An attacker could shut down cellular coverage in a given area by sending a large number of SMS messages to people who are in a given area. This is done using LBS on top of the SMS threat. Another what-if is the LBS virus threat. What if an attacker can disable, steal, or erase someone's phone when that person is in a particular place?

Another more real-life threat relating to smart phones is the fact that most of these devices have the ability to support a direct connection to the Internet. This means that these devices are out on the hacker-plagued Internet. The real threat comes from a hacker hacking into the device.

## 11.6  Wi-Fi Phones

Wi-Fi phones use the network for phone calls and require optimal bandwidth to operate. They are used to locate employees who work in a large area or who do not have a desk. They can also be used to roam from a Voice-over-Internet-Protocol (VoIP) network to a cellular network.

Wi-Fi phones have a major problem with regard to security in the fact that security processing takes time. The time used for security processing causes phones to have bad reception or drop calls. This has created another risk with Wi-Fi phones: either buy into a proprietary solution to increase the security processing of your network or lower the security to accommodate the Wi-Fi phones.

## 11.7  Chapter 11 Review Questions

1. What program is NOT in CE.NET but is in Pocket PC?
   a. Activesync
   b. Windows Media Player
   c. Outlook
   d. Internet Explorer

2. What was one of the first PIM devices?
   a. PLAM
   b. CE.NET
   c. Smart phone
   d. Zoomer

3. What is one of the biggest risks facing wireless devices?
   a. Theft
   b. Hacking
   c. Information loss
   d. Data corruption

4. Healthcare has seen a great increase in productivity due to which wireless end device?
   a. Laptop
   b. Tablet
   c. Smart phone
   d. Wi-Fi phone

5. Windows' first version of CE was based on _____.
   a. Windows NT4
   b. Windows 2000
   c. Windows 95
   d. Windows 98

6. What technology do most smart phones have that can be used to track them?
   a. CDMA
   b. CE.NET
   c. GPS
   d. PocketPC

7. When was the first computer virus written?
   a. 1980
   b. 1990
   c. 1983
   d. 1999

8. What security feature has Microsoft put into its Smartphone OS and PocketPC OS?
   a. Security filtering
   b. User action required to run an executable file
   c. Anti-virus
   d. Firewall

9. What makes securing a wireless end device more difficult?
   a. The large number of device types
   b. Multiple connection options (e.g., WLAN, PAN, cellular)
   c. Loose security
   d. Extra batteries

10. What does the acronym "LBS" stands for?
    a. Lucky basic security
    b. Location-based security
    c. Location-based services
    d. Location basic service

11. Which operating system was created from the ground up as a PDA, phone, and PIM combination?
    a. Smart phone
    b. RIM
    c. Symbian
    d. CE.NET

12. What is a major reason why handheld scanners have not gone 802.11g?
    a. Security
    b. Speed
    c. Battery life
    d. Chip set availability

# Chapter 12

# Wireless LAN Security

Wireless communications were always prone to security issues well before any of the 802.11 standards. Most people never thought about wireless security until the market responded with news, ads, and products to make the public aware of the dangers. What makes wireless such a security threat has to do with the fact that it is wireless. This means that data is transmitted over airspace and is susceptible to eavesdropping by anyone in a given area. Over the years, different types of encryption have been used to protect the data inside this transmission; however, this has not always been successful. This chapter walks through the history of wireless local area networking security, seeing how the industry went from welcoming wireless with open arms to fighting it day by day as people took it upon themselves to deploy it. This chapter looks at the history and then identifies most of the security methods available today. After reading this section, one will have a clear understanding of the methods in use in the past and present to protect wireless networks.

## 12.1 Wireless LAN Security History

Looking at the history of wireless networking, there are a number of interesting points, articles, attacks, and tools that were released over time. This section reveals how many of these wireless security methods were created, broken, and improved. This helps us understand how the market reacted to each new security solution and its subsequent exploitation.

On October 27, 2000, Jesse R. Walker published one of the first articles detailing the risks and issues regarding security in wireless 802.11b networking. In this article, he made comments about wireless networks being similar to data drops in the parking lot for anyone to use. He also pointed out a key point about increasing the key size of the WEP cipher during a time when the industry was pushing to increase the size of WEP keys from 40 bits to 128 bits. Walker pointed out that increasing the size of a WEP would not help against many of the attacks that were known at that time.

An article entitled "Your 802.11 Wireless Network Has No Clothes," published by the University of Maryland, was released on March 30, 2001. It contained details about a flaw in shared key authentication. This flaw took advantage of a technique widely known as plaintext cryptanalysis. This exploit worked by having access to a cleartext piece of data and its encrypted version. When comparing these the two, one can derive the key used to encrypt it. On a wireless network at that time, this key was used for most communications across these networks.

On July 16, 2003, Scot Fluhrer, Itsik Mantin, and Adi Shamir presented a paper titled "Weaknesses in the Key Scheduling Algorithm of RC4" at a security conference. This paper is one of the first papers to exploit the WEP encryption method itself. It stated mathematically how to break a WEP key, although it also states that the authors never tried it in a practical lab test. This paper was the main reasoning behind the wireless industry considering WEP unacceptable and in need of replacement.

After the release of "Weaknesses in the Key Scheduling Algorithm of RC4", members of AT&T decided to perform a real test on WEP encryption itself. This was because the author of the paper detailed how to do it without stating how they really did it. So the AT&T Lab members created a report, Labs Technical Report TD-4ZCPZZ, "Using the Fluhrer, Mantin, and Shamir Attack to Break WEP." This test was successful in breaking WEP encryption. Interestingly enough, the day this report was published the first WEP cracking tool, called WEPCrack, was also released. This tool was a Perl-based script tool with the ability to crack WEP keys. Five days later on August 17, 2001, Airsnort, a GUI WEP cracking tool, was released. For some reason, many members of the press acknowledged Airsnort as the first WEP cracking tool, when in actuality it was not.

On August 21, 2001, the "Weaknesses in the Key Scheduling Algorithm of RC4" paper was publicly released. After this paper was released, the AT&T folks had an issue with their lab results. It would be difficult to have a lab test results published before publishing the theory they were testing. This led to an update; in order to follow up with the "Weaknesses in the Key Scheduling Algorithm of RC4" public release, the AT&T testing paper was updated to reflect the correct date.

On February 6, 2002, The University of Maryland published an article entitled "An Initial Security Analysis of the IEEE 802.1x Standard." By this time, the industry's answer to the WEP security concerns was to move to an 802.1x security method. This article detailed many issues surrounding the 802.1x standard and the lack of two-way authentication. This gave hackers the ability to perform a man-in-the-middle attack on 802.1x wireless networks.

After the WEP flaw was released and tools were now considered mainstream, we started to see a number of networks being comprised, one of which involved a member of the press and a network security professional. They were driving around demonstrating a new trend called war driving. During this war drive, they went up and down a number of streets in Silicon Valley. Some of the networks they found included Sun Microsystems and Nortel Networks, both of which did not even have WEP enabled. Later, when the story hit the press, Sun Microsystems commented that the network was a test lab that was not connected to any part of its internal infrastructure. Nortel refused to comment on the press release.

On May 1, 2002, Best Buy had an encounter with wireless hackers. Best Buy was using wireless POS systems, which increased its ability to move its register's layout during peak purchasing times, such as Christmas. Some store POS systems were set up with the wireless services encryption enabled. This meant that when a customer used his Best Buy credit card, the number went across the airwaves to the corporate server for validation. Some wireless hacker sitting out in the parking lot was able to capture this traffic and get the numbers of multiple people's credit cards.

On September 22, 2002, Cisco released a response about the weakness of 802.1x cited by the University of Maryland; this article, entitled "Cisco Aironet Response to University of Maryland's Paper," stated that if one uses Cisco's proprietary method of authentication called LEAP, one would not have the problems pointed out in the previous University of Maryland article. This was because Cisco had two-way authentication between both the client and the access point. This meant not only was the client authenticating to the network, but the network was also authenticating to the client.

On November 13, 2003, Paul Timminsand and Adam Botby were arrested by the FBI outside a Lowe's home improvement store in Michigan. They were accused of "breaking into" a number of Lowe's stores across the Midwest. They were breaking in through the wireless networks and from there they were trying to hack into a mainframe computer located at Lowe's headquarters.

On November 27, 2003, a tool came about that could exploit LEAP. This tool, called anwrap, allowed a user to try multiple authentication attempts to the access point. This was easily combated using a password lockout policy, forcing an account to lock after a predefined number of unsuccessful attempts. This led to one of the latest tools called asleap. This tool was released on April 8, 2003, by Joshua Wright. It is also a LEAP attacking tool, but unlike the one previously released, this one has the ability to perform offline dictionary attacks on LEAP passwords. This tool prompted Cisco to abandon LEAP in favor of a new protocol called EAP-FAST.

## 12.2 Authentication

When connecting to a wireless network, one must perform some type of authentication. There are two main types of authentication per the current IEEE standards: share key authentication and open key authentication.

The next section will detail both authentication types, detailing how they operate and what security is present in their use.

### 12.2.1 Shared Key Authentication

Shared key authentication was created to be the more secure of the two types; however, as we will shortly see this actually became the less secure due to a small oversight in how it validates user keys. To understand shared key authentication the section below will explain how it works.

Shared key authentication works via a challenge response mechanism. To explore this process, one must first connect to the network. Having the client device send out a probe frame does this. This frame will look for available wireless networks and their connection settings. Once an access point "hears" a probe, it will respond with a probe response frame. This frame will identify all of its connection settings to the end device. In some cases, an end device will hear many responses from different access points in the area. To make sure that the end device connects only to the access point with the best signal, the probe response frame has a value for current signal strength. A client might hear multiple replies, although it will only connect to the access point with the highest signal strength value. Once the end client hears this and determines that it supports the same settings as the access point, the next portion (called authentication) takes place.

When the end device wants to authenticate, it sends an authentication response frame to the access point. This frame is evaluated; once the access point determines it is an authentication request, it will send a challenge packet back to the client. The challenge packet consists of a cleartext piece of data. The end device is required to encrypt this data with its WEP key and sends it back to the access point. Once this is done and the access point receives the packet, it checks it against what it has for the encrypted version of that packet. If the results match, the access

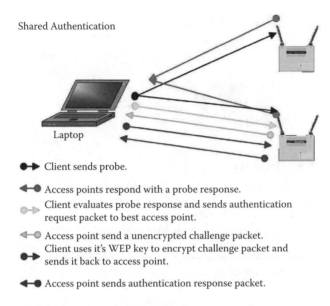

Shared Authentication

Laptop

●━▶ Client sends probe.

◀━● Access points respond with a probe response.

○━▷ Client evaluates probe response and sends authentication request packet to best access point.

◀━○ Access point send a unencrypted challenge packet.

●━▶ Client uses it's WEP key to encrypt challenge packet and sends it back to access point.

◀━● Access point sends authentication response packet.

**Figure 12.1   Shared key authentication.**

point will allow the end device onto the network. If the results do not match, the authentication fails and the end device is denied a connection. To get a full understanding of how this works, look at Figure 12.1, which outlines the connection and authentication process.

## 12.2.2  Open Key Authentication

Open key authentication was originally seen as less secure than shared key authentication. The intent was to make an open network, thus not requiring clients to have knowledge of the WEP key. As security became an increasingly visible issue, many vendors turned to the drawing board. This created a real problem: coming up with a solution that improves security while staying within the standard guidelines. These efforts led to the idea of using open authentication and, unlike before, this open authentication would require the use of a WEP key. When used, the WEP key was required to connect to the network. This worked because when one talked with the right WEP key, one's cyclic redundancy check (CRC) passed its test and the frame was allowed to access the network to it destination.

Looking at how open authentication works, one sees that the end device connected to the network as it did with shared key. It makes a probe request, listens to probe responses from multiple access points in the area, and then determines the best access point to make a connection with based on signal strength.

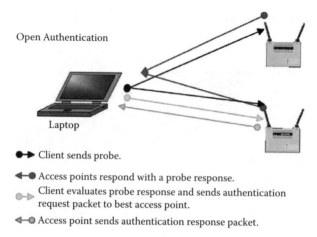

Open Authentication

Laptop

●—▶ Client sends probe.

◀—● Access points respond with a probe response.

●▸▸ Client evaluates probe response and sends authentication request packet to best access point.

◀—● Access point sends authentication response packet.

**Figure 12.2 Open key authentication.**

How do open and shared key authentication differ? Open authentication sends an authentication request but does not receive a challenge; instead, it is allowed to talk by default. When WEP is enabled, the process is slightly different. When the wireless client starts to talk, it automatically encrypts all the data with WEP encryption. When the access point hears data being sent, it decrypts the frames and forwards them. If the frames are encrypted with a different key than the access point, the decryption portion fails and the packet is dropped. To get a better understanding of how open authentication works, as well as how it differs from shared key authentication, look at Figure 12.2.

## 12.3 SSID

When one looks at security with respect to the SSID field, one sees that most networks, by default, broadcast this information to anyone who is listening. As more people started to take a deeper look into the security of wireless, they considered hiding the SSID in beacon frames. It was noticed that if the SSID was not broadcast, the existence of a wireless network could be somewhat masked. This masking would require the client to send a probe for any available wireless networks. In most IEEE wireless networks, the existence of the SSID is easily attainable with some sort of wireless sniffer. This is because the SSID is part of the process of connecting to a wireless network. Looking at Table 12.1, one can see that the SSID is a present in the header of a wireless probe request frame. This information can be read by any sniffing program, thus defeating any

**Table 12.1    SSID Sniffer Capture**

| | |
|---|---|
| 802.11 Beacon | FC=........,SN= 448,FN= 0,BI=100,SSID=,DS=11 |
| 802.11 Probe Req | FC=........,SN= 689,FN= 0,SSID=AEE |
| 802.11 Probe Req | FC=........,SN= 690,FN= 0,SSID=AEE |
| 802.11 Beacon | FC=........,SN= 449,FN= 0,BI=100,SSID=,DS=11 |
| 802.11 Probe Req | FC=........,SN= 691,FN= 0,SSID=AEE |
| 802.11 Probe Req | FC=........,SN= 692,FN= 0,SSID=AEE |
| 802.11 Probe Req | FC=........,SN= 693,FN= 0,SSID=AEE |
| 802.11 Probe Rsp | FC=........,SN= 451,FN= 0,BI=100,SSID=AEE,DS=11 |
| 802.11 Probe Rsp | FC=...R....,SN= 451,FN= 0,BI=100,SSID=AEE,DS=11 |
| 802.11 Probe Req | FC=........,SN= 694,FN= 0,SSID=AEE |
| 802.11 Probe Rsp | FC=........,SN= 452,FN= 0,BI=100,SSID=AEE,DS=11 |
| 802.11 Probe Req | FC=........,SN= 695,FN= 0,SSID=AEE |
| 802.11 Probe Rsp | FC=........,SN= 453,FN= 0,BI=100,SSID=AEE,DS=11 |

attempts to hide this identification information. Even with the SSID masked, every time a client wants to connect to a network, it will send all of its connections settings, including the SSID, out into the air as part of its probing process.

Several vendors have default SSIDs that they program into their equipment. This is one of the first avenues that a hacker will take when trying to exploit a wireless network. Many companies also use very simple SSIDs such as Wireless, WLAN, and BRIDGE. Table 12.2 outlines some of the SSIDs commonly used.

## 12.4  Wireless Security Basics

Before delving into all the different types encryption and advanced security methods, let us take a look at some basic things that can be done to create a secure wireless network. As shown, the SSID does not help security; however, it is still a good idea to prevent its broadcast. In addition, another important thing is to change the default usernames and passwords on all access points. Some even have default WEP keys that should be changed. Next, run open authentication with WEP enabled. Other recommended changes are to prevent DHCP and to segment the wireless network. One key goal would be to develop an information classification policy to educate everyone as to what should and should not cross the airwaves.

**Table 12.2   Default SSID Detail**

| | |
|---|---|
| ■ Cisco (All Aironet Access Points/Bridges) | |
| – SSID: | tsunami |
| ■ LINKSYS Product Families: | |
| – SSID: | linksys |
| ■ Netgear 802.11 DS products, ME102, and MA401Default | |
| – SSID: | Wireless |
| ■ SMC Access Point Family | |
| – SSID: | WLAN |
| ■ SMC2682W EZ-Connect Wireless Bridge | |
| – SSID: | BRIDGE Wave LAN Family: |
| – SSID: | "WaveLAN Network" |
| ■ Symbol AP41x1 and LA41x1/LA41x3 | |
| – SSID: | 101 |
| ■ TELETRONICS WL-Access Points | |
| – SSID: | Any |

## 12.5  Equivalent Privacy Standard (WEP)

The Wired Equivalent Privacy (WEP) standard was created to give wireless networks safety and security features similar to that of wired networks. WEP is defined as the optional cryptographic confidentiality mechanism used to provide data confidentiality that is subjectively equivalent to the confidentiality of a wired local area network (LAN) medium that does not employ cryptographic techniques to enhance privacy. This gives us the basic thought of how WEP was created and what goals it originally intended to meet. To meet these goals, wireless had to address the three tenets of information security: (1) confidentiality, (2) availability, and (3) integrity.

1. The fundamental goal of WEP is to prevent eavesdropping, which is confidentiality.
2. The second goal is to allow authorized access to a wireless network, which is availability.
3. The third goal is to prevent the tampering of any wireless communication, which is integrity.

To better understand WEP, one needs to look at it more closely. The WEP protocol is used to encrypt data from a wireless client to an access point. This means the data will travel unencrypted inside the wired network. The WEP protocol is based on RSA Securities' RC4 stream cipher. This cipher is applied to the body of each frame and the CRC. There are two levels of WEP commonly available: (1) one based on a 40-bit encryption key and 24-bit initialization vector, which equals 64 bits; and (2) one based on a 104-bit encryption key and 24-bit initialization vector, which equals 128 bits.

This protocol has been plagued with issues since its inception. A magnitude of exploits, poor design elements, and general key management problems have made WEP a very insufficient security mechanism. The details regarding the problems of WEP are explored in Chapter 13. This section, however, discusses how WEP operates. One of the original functions of WEP was to have the encryption unable to be affected by loss of the frame due to interference. What this means is when one sends data across the air and loses the frame, there would be no loss to the previous frame. With newer security methods and older wired secured methods, it is common for subsequent packets to have an encryption dependency on the next or previous frame.

## *12.5.1 WEP Encryption Process*

As learned in the previous section, WEP uses an RC4 stream cipher to encrypt wireless data. In performing the WEP encryption process, a number of steps are performed. The first step is to generate a seed value. This seed value is used to start the keying process. This value can be referred to as a key schedule or as the seed value. No matter what this value is called, it is considered the WEP key. After this value is defined, it must then be entered into the access point. To ensure that the client can receive and decrypt the transmission, the seed value or WEP key must be entered on to each client. This will allow a WEP encrypted conversation to occur. This value consists of a 26-digit hexadecimal number.

This value is not used alone to create a WEP encrypted data stream; a technique to randomize the key is applied as well. This technique uses a 24-bit initialization vector (IV) that is created on a frame-by-frame basis. The technique on which the IV is created differs between vendors. The WEP standard that is outlined inside 802.11b states the IV size and requires that the IV change on a frame-by-frame basis. Outside of this, there are no requirements in the standard defining how to increment or randomize the IV sequence. As one progresses through this book, one will see that the missing definition of an IV sequence is one place where the WEP protocol has a major security issue. Once the IV and WEP key are together, they can be used to encrypt the frame. When the data is ready for

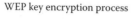

WEP key encryption process

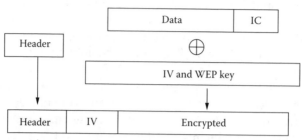

**Figure 12.3   WEP process detail.**

transmission, the WEP key and the IV are combined; then, using the RC4 cipher, the key and IV are XORed with the data and IVC to create the encrypted frame. Next, a copy of the same IV is put into the frame header as cleartext. The last step is to send the packet.

Once the other end receives the frame, the IV is picked out of the frame header and applied to the predefined seed value to produce the same session-based WEP key that was used to encrypt the packet. The same RC4 encryption process is performed in reverse, allowing the encrypted text to turn from ciphertext to plaintext. Once this operation is complete, the CRC is removed and applied to the data to make sure that it was not corrupted in transit.

Figure 12.3 shows how the encryption takes place. As one can see, the data and the integrity check are put together linearly as well as the WEP key and the IV. The WEP key and IV are fed to the pseudo-random number generator. The output of this generates a key stream equal to the length of the frame payload plus the CRC. The sender then performs an XOR operation on the key stream and the data, creating an encrypted data stream. The next step involves placing the IV inside the packet header and forwarding the packet to the sender.

# 12.6  802.1x

Both the IEEE and ANSI organizations approved the 802.1x standard. On June 14, 2001, the IEEE approved the standards; four months later on October 25, 2001, the American National Standards Institute (ANSI) approved it as well. The 802.1x standard was designed for port base authentication for all IEEE 802 networks. This means it will work across Ethernet, FDDI, token ring, wireless, and many other 802 networking standards.

One thing people tend to become confused about is that 802.1x is in no way any type of encryption or cipher. All the encryption takes place outside the 802.1x standard. For example, on a wireless network, the EAP would use one of its various methods of encryption for authentication. After the user authenticates to the wireless network, they may start a conversation using WEP, TKIP, AES, or one of the many other standard wireless encryption schemes. When looking at the 802.1x standard, at its most basic view one sees actually what it was intended for, port-based authentication. This means the standard takes the authentication request, decides if it is or is not allowed onto the network, and then grants, or revokes, access.

Many parts of how 802.1x works are within other standards such as EAP and RADIUS. The 802.1x standard is just a mechanism that denies all traffic except EAP packets from accessing the network. Once the EAP says it is OK for the device to access the network, the 802.1x protocol tells the switch or access point to allow user traffic. This is accomplished by having the network port or, in a wireless situation, each client connection in one of two port states. These states are controlled and uncontrolled.

Figure 12.4 reveals the three main designations called out by the 802.1x standard. Each of them has specific rules and functions. The standard was written to incorporate a large amount of different equipment; the names of these functions remain somewhat generic. As one can see from Figure 12.4, the 802.1x protocol leverages two other standards. From the supplicant to the authenticator, the standard is EAP. From the authenticator to the authentication server, the protocol is RADIUS. The 802.1x protocol takes EAP requests, sends them to a RADIUS server, and waits for an answer. Once this answer is received, it will allow or deny access to the network.

Looking at each of the parts of Figure 12.4, one can explore the key roles in the 802.1x standard. The authentication server, authenticator, and supplicant are three main roles of any 802.1x exchange. They each perform specific roles in processing the authentication exchange and allowing correctly authenticated devices or users onto the network.

## 12.6.1 Authentication Server

The authentication server provides the access granting and access rejecting features. It does this by receiving an access request from the authenticator. When the authentication server hears a request, it will validate it and return a message granting or rejecting access back to the authenticator. This is the back end of the 802.1x standard and, per the standard, the operation of this server is defined in another standard (i.e., RADIUS).

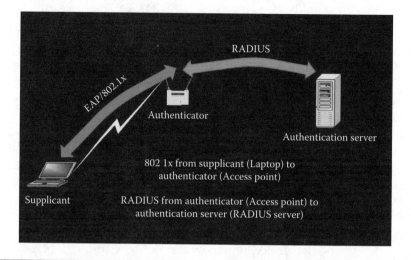

**Figure 12.4   802.1x Overview.**

## 12.6.2 Authenticator

The authenticator is the first piece of network electronics that an 802.1x device will attempt connection. In this example, it is a wireless access point, although it can be anything providing access to the network. The device's role is to let only EAP packets pass through and then wait for an answer from the authentication server. Once the authentication server responds with an accept or reject message, the authenticator acts appropriately. If the message is returned, it is a reject message and it will continue to block traffic until the result is an access accept. When the accept response comes from the authentication server, the authenticator then allows the supplicant the ability to access the network.

## 12.6.3 Supplicant

The supplicant is the device that wants to connect to the 802.1x network. This can be a computer, laptop, PDA, or any other device with a network interface card. When the supplicant connects to the network, it must go through the authenticator. This authenticator only allows the supplicant to pass EAP request traffic destined for the authentication server. This EAP traffic is the user's or device's authentication credentials. Once the authentication server determines that the user or device is allowed on the network, it will send an access-granting message.

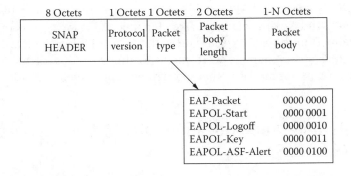

**Figure 12.5  EAPOL diagram.**

## 12.6.4 Extensive Authentication Protocol over Local Area Network (EAPOL)

Extensive Authentication Protocol over Local Area Network (EAPOL) is part of EAP, although it is outlined inside the 802.1x standard. Because of this, it is located in the 802.1x section in this book. This is because the 802.1x standard allows certain EAP message types to pass through an authenticator to the supplicant. To define what messages are allowed to pass through, each message type and frame format had to be included inside the 802.1x standard. The EAPOL standard calls out the process and frame structure used to send traffic from the authenticator to the supplicant. This traffic is outlined with six frame types. This means that only these six frame types are allowed to pass through an access point to a client. The IEEE created room for more, although the current standard only outlines six. Figure 12.5 shows each frame type, along with what value is used to identify the frame.

- *EAPOL-Packet.* This frame type is used to identify the packet as an EAP packet.
- *EAPOL-Start.* This frame type is used to begin an EAP conversation or an 802.1x authentication.
- *EAPOL-Logoff.* This EAPOL frame is used to end an EAP conversation or an 802.1x authentication.
- *EAPOL-Key.* This EAPOL frame is one of the most security-related frames. It is used to exchange keying information between the authenticator and the supplicant.
- *EAPOL-Encapsulated-ASF-Alert.* This is an EAPOL frame used to carry SNMP trap information out a non-802.1x authenticated port.

The most involved EAPOL type is the EAPOL-Key frame. This frame is used to send keying material, like dynamic WEP keys. The only key frame defined in the 802.1x standard is the RC4 WEP key. In 802.11i, one will see that some changes were made to the operation of the EAPOL-Key frame to accommodate other encryption cipher types outside of RC4.

## 12.7  Remote Authentication Dial-In User Service (RADIUS)

RADIUS is an acronym for Remote Authentication Dial-In User Service, a protocol used in network environments for authentication, authorization, and accounting. RADIUS can run across many types of devices, such as routers, servers, switches, modems, VPN concentrators, or any other type of RADIUS-compliant device. The protocol works by creating an encrypted tunnel between the network device and the RADIUS server. This tunnel is used for sending all the Authentication Authorization Accounting (AAA) information about who a user is, where the user is allowed to go, and where the user actually did go. To start this encrypted tunnel, a phrase or password called the shared secret is needed. The shared secret is located on the RADIUS participating network device and the RADIUS server. Once the shared secret is correctly set up, secure communication can take place.

One of the benefits of RADIUS is the use of a common database of users to provide these AAA services across multiple device types. The database that RADIUS uses for storing usernames and passwords can be set up to point to many different types of directories. This means RADIUS can use most existing directory structures, such as Microsoft Active Directory (MS-AD), Novell Network Directory System (NDS), Lightweight Directory Access Protocol (LDAP), and many other common directory types.

This protocol allows administrators to centrally locate and administer user access and accounting for all network equipment as well as remote access. RADIUS prevents many of the headaches associated with properly removing access to network equipment when employees are terminated. Once an organization has deployed RADIUS, user access could easily be removed in the event of a termination. This was unlike in the old days when administrators would have to manually change usernames and passwords on all network equipment.

The RADIUS protocol specifications are currently defined in RFC 2865 and RFC 2866. RFC 2865 focuses on the access portion of RADIUS, allowing user access into devices or onto the network. RFC 2866 focuses on the accounting portion of RADIUS, allowing administrators to track changes and access to network devices as well as general access to the network itself. The original RFC numbers were 2138 and 2139; these were updated to address a number of security-related concerns. Another major reason for this update was to change the UDP port number of RADIUS from the original port numbers of 1645 and 1646 to 1812 and 1813. The protocol was changed because the UDP port number of 1645 was already designated by IANA for the datametrics service and not for RADIUS. This is why most RADIUS servers support all four of these ports by default.

Now that we have had a look into the primary usage and history of RADIUS, one can now look at how RADIUS relates to wireless. First, RADIUS could be used as an access method to administer the access point. This is similar to how it would be used to administer routers or switches. The access point and the RADIUS server would have a shared secret and that would be used to set up an encrypted channel that can carry user authentication traffic.

Another approach is to use RADIUS, as explained in the 802.1x standard, as a back-end user authenticating mechanism. RADIUS itself provides this feature, so the 802.1x standard used that instead of creating its own authenticating mechanism. With this scenario, the access point would need to be set up correctly with the RADIUS server's shared secret and the access point would keep track of the user's request to enter the network. This means the user would only negotiate its authentication with the access point, not the RADIUS server. This is similar to how RADIUS would be set up on a remote access device. A user would ask to enter the network; the user would then be prompted by the network device to provide some type of authentication. Once the user provided authentication, the network equipment would verify it against the RADIUS server user database. If the credentials are correct, the user would be allowed onto the network; if the credentials are not correct, the user would be denied access.

RADIUS has only four types of packets for authentication and there are other packet types for accounting. However, this section only focuses on the authentication packets. The four types are as follows:

1. *Access-Request.* This packet allows the RADIUS sequence to take place.
2. *Access-Accept.* This packet informs the RADIUS client that the authentication provided to it was correct.
3. *Access-Reject.* This packet informs the RADIUS client that the authentication provided to it was incorrect.
4. *Access-Challenge.* This packet is used to challenge a RADIUS client for its authentication credentials.

RADIUS packet formats

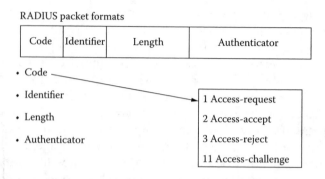

**Figure 12.6   RADIUS frame format.**

Looking more closely at the RADIUS packet in Figure 12.6, one can see that all four packet types consist of the same packet format. They are identified by the code field. This field has a number of available codes that could be used; however, the only ones addressed here are the Access-Request identified by a 1, the Access-Accept identified by a 2, the Access-Reject identified by a 3, and the Access-Challenge identified by a 1. The next field (the Identifier) is used to match requests and responses to each other. This ensures that multiple RADIUS conversations do not get mixed up as to what messages go to what device. The Length field is used to identify the length of the packet; because the RADIUS packet can have up 2000 attributes inside it, a mechanism is needed to measure the packet length. The last field, the Authenticator field, is the field in which the password is protected; this password is protected by a hashing mechanism.

RADIUS has a lot of complexity. According to RFC 2865, each RADIUS packet can be up to 4096 bytes, allowing 2000 attributes into a single packet. Setting up RADIUS can be easy; however, depending on which vendor solution one decides on, there could be added complexity. Once thing to note is that there is no security complexity; time has shown that if security is too complex, it will be avoided or not installed.

When looking at the RADIUS server, as well as any authentication servers for that matter, the details surrounding protecting the server itself are often forgotten. This stems from network people building and administrating networks — not servers. This leaves the server that RADIUS sits on exposed. For example, if the network people build and administrate networks, they most often do not have the skills to secure servers. One can have the most secure network in the world; however, if someone can easily hack into one's authentication server, then one's whole network is compromised.

# 12.8 Extensible Authentication Protocol (EAP)

Extensible Authentication Protocol (EAP) is a standard method of performing authentication to gain access to a network. When Password Authentication Protocol (PAP) first came out, security issues quickly made it a less desirable authentication method. After that, the Challenge Handshake Authentication Protocol (CHAP) came out and this also quickly became plagued with security issues. The industry decided it was easier to make an authentication protocol act the same way no matter how or what type of authentication validation took place. This meant that for the first time a protocol could be inserted into products and software that allowed for passwords, tokens, or biometrics without having to write any extra code to support the different methods. This is how and why EAP was created. To use EAP, one must specify inside the type field what kind of authentication one is going to use. This allows one to use EAP for password, tokens, and other authentication types. EAP can adapt to security issues and changes by leveraging different methods of authentication. EAP also is able to address new and always-improving authentication techniques without having to make any changes to EAP supporting equipment.

When EAP was created, a need for PPP compatibility was required. This helped ensure that a large base of equipment could handle EAP without major modifications. To get this compatibility, EAP was included as a Point-to-Point Protocol (PPP) type inside a PPP packet itself. This allowed for any device supporting PPP to be able to support EAP. EAP remained this way through RFC 2284. As EAP matured and required tighter integration with the 802.1x standard, its placement inside PPP was evaluated. The result of that evaluation is RFC 3748. Quoted below from RFC 3748 is the reasoning behind the changes to EAP that were needed to support the 802.1x standard. "The IEEE 802 encapsulation of EAP does not involve PPP, and IEEE 802.1X does not include support for link or network layer negotiations. As a result, within IEEE 802.1X it is not possible to negotiate non-EAP authentication mechanisms, such as PAP or CHAP [RFC1994]."

EAP is defined in a PPP packet by the Hex code C227. PPP does not require authentication, although it does support it. As one can see from Figure 12.7, the EAP process starts by telling the device that the PPP authentication method is taking place and that method is EAP.

Looking at the EAP portion of the packet itself, one can see that it is made up of four main fields: code, identifier, length, and data. These fields are used to identify what the packet function is, how long it is, and which current EAP negotiation the packet belongs to. Each of the sections in the EAP packet is identified in Figure 12.8, outlining each byte that makes up an EAP frame.

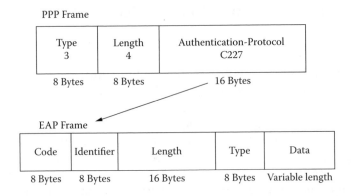

**Figure 12.7    PPP EAP frame format.**

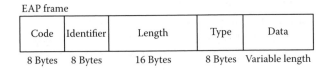

**Figure 12.8    EAP frame detail.**

- *Code field.* The Code field has a single octet and currently allows for four types of codes. The Request code makes an initial request; the Response code informs the requestor that their request was made. These codes are differentiated by the Code field being set to 1 or 2. The next set is the Success code, which validates that the authentication took place. It can inform the other party with a succeeded or failure code. These packets are distinguished from the other using a 3 in the Code field to indicate a success and a 4 to indicate a failure. When the RFC was updated to 3748, a note was placed in the Success section dealing with the arrived order of success packets. What this means is that if a client receives a Success frame before any Response frames during the EAP process, it will drop the frame. This was created to prevent attempts to circumvent the security of EAP.
- *Identifier field.* This section is also a single octet long. The function of the identifier attribute is to match requests to responses. This is used to distinguish when multiple clients authenticate with EAP at the same time. In this scenario, each client has a unique identifier that keeps its success or failure codes matched to each individual request.

- *Length field.* This field is two octets long and indicates the length of the EAP packet. The length includes the code, identifier, length, type, and, of course, the actual data portion.
- *Type field.* The Type field has one octet and identifies the structure of an EAP Request or Response packet. This is where each EAP type allows for many kinds of authentication to take place. The EAP type is discussed later when we look at a number of the common EAP types that relate to wireless. This section has four main type codes to control the transmission of EAP credentials to the authentication server. These four types of type codes are identity, notification, NAK, and MD5-Challenge. The identity code is used to query the identity of a client and to start the process of identifying what type of EAP method will be used to authenticate. The notification code is used to notify the client of something or to get them to perform some type of action before continuing. The next code is NAK, which is valid only in response messages. This code is used for informing the client that the authentication type is not valid. There is also the ability to send an extended NAK, which details what EAP types the authenticator would accept.
- *Data field.* This section has the actual data that makes up the EAP request. Inside this part is where a user would provide the authentication portion. If EAP is using MD5, that would mean the user would have to submit a username and password to the access point. This information would be hashed and then located inside the data field of the EAP packet.

One of the main points in using EAP is the ability to leverage multiple types of authentication mechanisms. This has helped EAP from becoming obsolete due to security vulnerabilities or protocol weaknesses. The ability to use multiple authentication types is located in the Type field on an EAP packet. The original standard as well as the new RFC 3748 only lists three main EAP types. These types are MD5 Challenge, One Time Password (OTP), and Generic Token Card (GTC). Today there are a number of different EAP types, some of which are vendor specific, some detailed in the EAP standard, and some detailed within their own standard documents. Others are industry standards on their own, detailed by IETF documents or RFC documents. In the subsections below, the most widely used wireless-related EAP types are detailed and examined.

## 12.8.1 EAP-MD5

Extensible Authentication Protocol-Message Digest version 5 (EAP-MD5) is one of the most limited EAP types included in the EAP RFC. This version

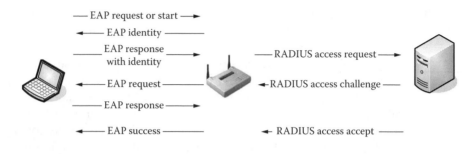

**Figure 12.9 EAP-MD5 detail.**

uses an MD hashing algorithm to validate user credentials. Some of the other types of EAP methods create encrypted tunnels and then inside these tunnels they perform EAP-MD5 validation. One of the requirements of EAP MD5 is a shared secret. This secret must be shared out of band so two parties know the same secret. That secret is then used to encrypt a challenge to verify that the other party has the same secret.

Looking at Figure 12.9 one can see how EAP-MD5 works. The first step is to have the client send a request packet to the access point. Before that, of course, the shared secret must be exchanged out of band. Once the access point hears this request, it will respond, asking the client to provide its identity. The identity would state what EAP type they are both going to use. The RFC 3748 for EAP allows other identification items to be located in the EAP identity message, although the primary function of the EAP Identity field is to inform the authentication server of what type of EAP method the client will be using. Contrary to what the name implies, the Identity field should not be used for user identity — only for EAP type identity.

Continuing with the example, the EAP type inside the EAP identity frame section would be set to MD5. Once the access point receives this identity code inside the frame of message 2, it would forward it to the authentication server. The authentication server would next send out a challenge to the client. This challenge is unencrypted and sent in cleartext. Once the client receives the challenge, it will use the shared secret to encrypt the challenge and send it back to the authentication server. Once the authentication server receives the challenge, it will perform the same operation that the client did. It will validate that the hash it has come up with matches the one that the client sent. If this is true, an EAP success message is sent; if for some reason something was not correct, then the authentication server would send a failure message with a reason code. This reason code would help to understand why the EAP frame was not accepted.

## 12.8.2 EAP-TLS

The Extensible Authentication Protocol-Transport Layer Security (EAP-TLS) method is described in RFC 2716 and was created by Microsoft in October 1999. The RFC was built on the following RFCs: RFC 2284 for PPP and RFC 2246 for TLS. TLS came about from the older SSL protocol. Netscape created Secure Socket Layers (SSL) and used it for secure Web browsing. Once the Internet took off, updates to SSL were required. In 1996, the IEEE created TLS based on Netscape's SSL and Microsoft's Private Communications Technology (PCT). This EAP method uses certificates to authenticate users and requires certificates at both the server and client end. This is where TLS plays into this standard; TLS already created a good way to perform the needed certificate validation steps. This particular EAP method is one of the strongest, although it prevents usage unless one is accessing the network from a computer with one's client certificate already installed.

Setting up a wireless network with 802.1x and EAP-TLS requires some up-front work and planning. First, one must have a certificate authority (CA); this server will function as the distributor of both client and server certificates. Also needed would be an AAA server that supports EAP-TLS type. Finally, one needs a client that can support this EAP type. Once all the pieces are in place, the next challenge of correctly configuring each piece to interact is necessary.

The EAP-TLS method works similar to the other EAP methods. To understand it, we present a high-level overview of the authentication process using this EAP method (Figure 12.10). It starts with a request packet originating from the client through the authenticator bound for the authentication server. Once this request is received, the authentication server will send back a response to the client asking for its identity. Next, the client will provide identity by requesting the use of EAP-TLS. Once this is received, the authentication server will send its public key certificate to the client. Once the client gets this, it will respond with its public key and a secure channel will be set up. This is very similar to how a secured Web page works, although in this case a client validation takes place. Now that both ends have authenticated to each other, another EAP process can take place inside the EAP-TLS tunnel to allow secure authentication to take place. For example, an EAP-MD5 exchange can take place inside the encrypted tunnel, allowing the weaker authentication to be secure because it is taking place inside an encrypted tunnel.

## 12.8.3 EAP-TTLS

The Extensible Authentication Protocol-Tunnel Transport Layer Security (EAP-TTLS) is an IETF draft document created by Funk Software Inc. The most recent version is named draft-ietf-pppext-eap-ttls-05.txt and was

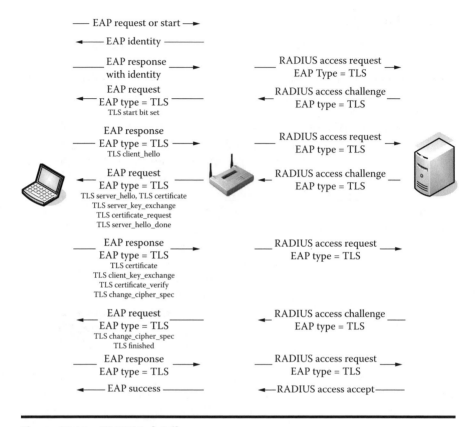

**Figure 12.10   EAP-TLS detail.**

created in July 2004. The reasoning behind creating a new EAP type was based on an opportunity Funk Software saw in the market. This opportunity was a need to support older devices that were not able to perform the new authentication types. This gave Funk the idea to write an EAP type that allowed for secure communication of credentials, along with the ability to allow legacy authentication types. This EAP type has helped Funk offer secure wireless solutions for older client equipment.

The EAP-TTLS method (Figure 12.11) works by taking the authentication and protecting it inside a TLS tunnel; however, unlike EAP-TLS, EAP-TTLS can use authentication types outside of EAP for the client authentication portion. Some older methods of authentication supported are PAP, CHAP, MS-CHAP, MS-CHAP-V2, and many others. Looking at how this works, one sees the main EAP process discussed thus far. The client makes a request with a request packet. This packet originates from the client through the authenticator bound for the authentication server. Once this request is received, the authentication server will send back a response to the client asking for its identity. Next, the client will provide identity

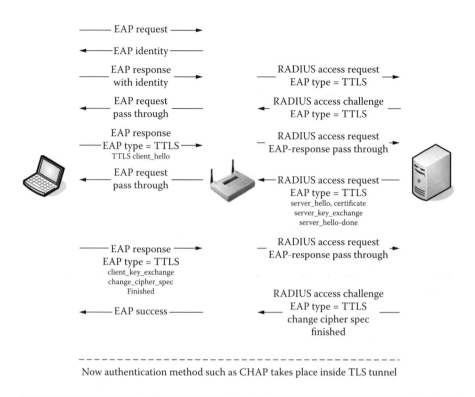

EAP request ——▶

◀—— EAP identity ——

EAP response
with identity ——▶

RADIUS access request
EAP type = TTLS ——▶

◀—— EAP request
pass through

◀ RADIUS access challenge
EAP type = TTLS

EAP response
—— EAP type = TTLS ——▶
TTLS client_hello

RADIUS access request
EAP-response pass through ——▶

◀—— EAP request
pass through

◀—— RADIUS access request ——
EAP type = TTLS
server_hello, certificate
server_key_exchange
server_hello-done

—— EAP response ——▶
EAP type = TTLS
client_key_exchange
change_cipher_spec
Finished

RADIUS access request
EAP-response pass through ——▶

RADIUS access challenge
◀—— EAP success ——

◀ EAP type = TTLS
change cipher spec
finished

Now authentication method such as CHAP takes place inside TLS tunnel

**Figure 12.11   EAP-TTLS detail.**

by requesting the usage of EAP-TLS. Once this is received, the authentication server will send its public key certificate to the client. An optional configuration is to have the client send a certificate to validate that it is the correct client. Our example will not have the client using a certificate because the whole point of EAP-TTLS is to leverage old equipment, which cannot use advanced authentication methods such as Public Key Infrastructure (PKI). If one is going to use or already has PKI, EAP-TLS would be the better option. Once the client receives the certificate from the authentication server, it will use EAP-TTLS to set up a secure channel between him through the authenticator to the authentication sever. This is done using TLS and the public key of the authentication server. These are almost the same steps that any client would take in accessing a Web page that is protected by TLS or HTTPS. Once this is complete, a number of other standard types of authentication can take place inside this TLS tunnel. This is where EAP-TLS and EAP-TTLS defer in EAP-TLS all that could be done at this point is to perform another type of EAP such as EAP-MD5 inside the already set-up EAP channel. With EAP-TTLS you can use other forms of authentication that are not possible on device that do not support EAP.

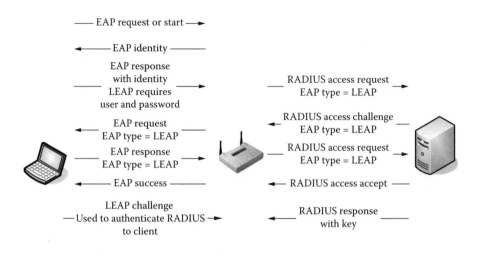

**Figure 12.12    LEAP detail.**

## *12.8.4 LEAP*

Lightweight Extensible Authentication Protocol (LEAP) is a Cisco Systems proprietary protocol. Cisco did release the source code for vendors that wanted to incorporate LEAP into their wireless adapters. The list of vendors includes D-Link, Dell, SMC, 3Com, and Apple. The code for LEAP is still considered Cisco Systems' intellectual property and is available for use only under a non-disclosure agreement (NDA).

LEAP works somewhat differently than the main EAP types discussed thus far (Figure 12.12). The EAP process starts out with the request from the client directed toward the access point. When the access point hears this, it will send the response and ask for identity. Most other EAP types just ask for the client EAP type; LEAP will respond with a username in the Identity field. Once the access point hears this, it will forward the packet to the authentication server or RADIUS in most wireless cases. The server will then send a challenge to the client through the access point.

The challenge is made up of an 8-byte message. When the client hears this message, it will perform a number of steps in order to respond. First, it will use MD4 hashing, which will produce a 16-byte hash of the password. This hash will have the last 5 bytes padded with nulls. This will create a total 21-byte hash. This hash will be broken up into 7-byte chunks and each chunk will go through a DES encryption process. Each process will use the 7-byte chunks as a key and the original 8 bytes as the plaintext. This will result in a number of DES operations, creating a ciphertext that is 24 bytes long. This text will be sent to the authentication server where it will do the same operation in reverse and derive the original password. Once it has this password, it will make sure that the passwords match that particular user. If a match is found, the authentication server will send an access-accept message to the access point. If not, the authentication server will send an access-reject message to the access point. Once the access point hears either of these messages, it will send an EAP failure or EAP success message to the client.

## 12.8.5 PEAP

Protected Extensible Authentication Protocol (PEAP) was created as a joint effort between RAS, Microsoft, and Cisco Systems. Currently, PEAP is in an IETF draft called draft-josefsson-pppext-eap-tls-eap-08.txt, last updated in July 2004. Because of the fact that it still exists in draft form, updates may change its version number or document name. PEAP was a move by the industry to make a single EAP method that multiple vendors could share. The three vendors that created the standard implemented it each in their own way; this made Microsoft and Cisco versions of PEAP different and not interoperable. This has been slowly working itself out of the system, although do not expect an easy integration when trying to use PEAP methods from Microsoft and Cisco interchangeably.

One of the main advantages of PEAP is the ability to have a strong EAP type that does not require client certificates, such as EAP-TLS. PEAP works similar to EAP-TLS by creating an encrypted tunnel with TLS and then performing another EAP method inside this encrypted tunnel. Unlike EAP-TLS, when PEAP performs this process, it does not validate a client certificate. This is where Cisco and Microsoft differ; each of them uses a different method after the TLS connection is created.

The PEAP process (Figure 12.13) starts similar to the other EAP processes with a request packet sent to the access point. Once the access point hears this request, it will respond with a response packet. This response will also ask the client for its identity. The identity will be the EAP type; in this case, one will see EAP-Type=PEAP field in the frame. This will instruct the authentication server that the EAP method being used will be PEAP. The traffic will just pass through the access point, which can also be called the authenticator. Once the EAP method has been determined, a TLS tunnel is set up using the authentication server's certificate to establish the TLS tunnel. Once a TLS tunnel is created, a new EAP process takes place inside the tunnel to authenticate the client. If one is using the Microsoft PEAP version, one will be using MSCHAPv2 as an EAP type. This EAP method is defined in the IETF document called draft-kamath-pppext-eap-mschapv2-01.txt, the latest version of which was posted in April 2004.

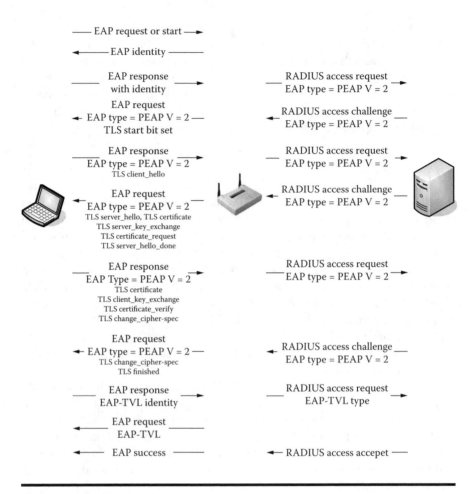

—— EAP request or start ——►

◄———— EAP identity ————

EAP response
with identity ——►

RADIUS access request
EAP type = PEAP V = 2 ——►

EAP request
◄— EAP type = PEAP V = 2 —
TLS start bit set

◄—— RADIUS access challenge ——
EAP type = PEAP V = 2

EAP response
EAP type = PEAP V = 2 ——►
TLS client_hello

RADIUS access request
EAP type = PEAP V = 2 ——►

EAP request
◄— EAP type = PEAP V = 2 —
TLS server_hello, TLS certificate
TLS server_key_exchange
TLS certificate_request
TLS server_hello_done

◄—— RADIUS access challenge ——
EAP type = PEAP V = 2

EAP response
EAP Type = PEAP V = 2 ——►
TLS certificate
TLS client_key_exchange
TLS certificate_verify
TLS change_cipher-spec

RADIUS access request
EAP type = PEAP V = 2 ——►

EAP request
◄— EAP type = PEAP V = 2 —
TLS change_cipher-spec
TLS finished

◄—— RADIUS access challenge ——
EAP type = PEAP V = 2

EAP response
EAP-TVL identity ——►

RADIUS access request
EAP-TVL type ——►

◄—— EAP request
EAP-TVL

◄—— EAP success ——

◄— RADIUS access accepet —

**Figure 12.13   PEAP.**

## *12.8.6 EAP-FAST*

Extensible Authentication Protocol-Flexible Authentication via Secure Tunneling (EAP-FAST) is an IETF document created by Cisco Systems in February 2004. The current document is named draft-cam-winget-eap-fast-00.txt and is located on the IETF Web site. Cisco had some security flaws released on their LEAP method and instead of fixing them, they abandoned the proprietary standard and created this EAP method. This EAP method supports a fast roaming time compared to the other EAP standards. This timing was a critical requirement for many companies that have Wi-Fi phones or some time-sensitive application. This made Cisco Systems in need of a fast, secure EAP method for wireless authentication.

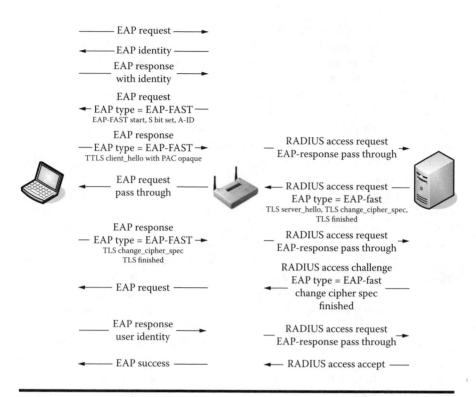

**Figure 12.14    EAP-FAST.**

The EAP-FAST method (Figure 12.14) works by having what is called a protected access credential (PAC), which is similar to a certificate. The PAC is a 32-octet key that is used to set up an encrypted tunnel between the client and the authentication server. One unique thing about this key is its ability to have the keys dynamically allocated. This PAC dynamic allocation process is called auto-provisioning, phase 0, or in-band provisioning. Whatever one decides to call it, it works by setting up a TLS connection and passing MSCHAP V2 credentials to access the network. To note is that the TLS session is only authenticated by the server's public key. After the initial network connection is made, a new PAC can be sent down this encrypted channel. Cisco does not recommend this approach if avoidable; however, it does allow for an easy upgrade from LEAP to EAP-FAST. The other approach is to use out-of-band key exchange or good old sneaker net. To do this, just key in the same key at both the user end and the server end. To get around any key management problems, the PAC has the ability to change at specified intervals once a secure connection has been established.

Looking more closely at the PAC, one can see that it is made up of three main portions: (1) the master key, (2) the opaque, and (3) the PAC info. These three parts each have their own functions in supporting EAP-FAST. The master key is generated randomly by the authentication server; this generation is done to produce strong entropy. After the key is generated, it is sent to the client. This key has the authority ID bound to it as well. The second portion of the PAC looks at the opaque, the authority ID, and key lifetime that are encrypted by the client's master key. The final portion of the PAC is the info section, which has the authority ID and can have the key lifetime. This field is used to difference multiple PACs on a client to support connection to multiple networks using multiple sets of credentials.

Now that the PAC has been created or keyed in, the EAP process can start. As with the other EAP processes, this one starts in the same manner with the request packet from the client. The access point answers the request and a response is sent back requesting identity. The client then sends its identity, which is not the client's identity but the actual EAP type identity. This identity request just requests the usage of EAP-FAST. Once the access point hears this identity request, it will respond with an EAP request to send the client's PAC. This message with the PAC will be passed on to the authentication server. The authentication server will then send an Authority ID, which is a field inside the PAC. This field identifies the correct PAC from a client PAC list. Once the client hears the authority ID and matches that against the correct PAC, it can start to create a TLS tunnel based on the Master key for that PAC. This will create a tunnel that can be used to perform further authentication.

## 12.9  Wi-Fi Protected Access (WPA)

Wi-Fi Protected Access (WPA) has an interesting history in relation to how it became a standard. When the security of WEP was broken, the industry turned to the IEEE to fix it. The IEEE said it would create the 802.11i wireless security standard. This standard dragged on and was very slow-moving. As it took longer and longer to ratify, wireless device sales declined. This decline was due to the lack of a standard secure wireless networking method. With this all-so-needed standard lacking, the wireless manufacturers started to push the IEEE and other standard boards to ratify something so they could produce secure standard products. With the pushback of the 802.11i release date, the Wi-Fi Alliance decided that it would create a subset 802.11i standard called WPA. The Wi-Fi Alliance created WPA by leveraging what the 802.11i task group had already done and formalized it into WPA. This meant that any major changes to the 802.11i standard would influence future versions of WPA. This was seen with WPA and WPA2. Today, with 802.11i complete, the use of WPA has greatly decreased.

The WPA standard supports two methods of authentication and key management. The first one is EAP authentication with the 802.1x standard. This method works through the use of the 802.1x protocol and a back-end authentication server. It leverages EAP for in-air authentication and RADIUS for back-end authentication. This method is the more secure of the two and provides the lowest amount of end-client administration. The next available option is to use preshared keys. This option requires that a key be applied to the devices and the wireless access points. This also means that everything has the same password entered. To combat someone using this key to eavesdrop on others' conversations, WPA uses a method that creates a unique session key for each device. This is done by having a preshared key called the group master key (GMK) that drives a pair transient key (PTK). How this works is explained in the section on 802.11i. This second solution was added to WPA for home and small office support. In a house or small office, one is unlikely to have an authentication server such as RADIUS. A PSK is a 256-bit number or a passphrase that is 8 to 63 bytes long.

WPA does support TKIP and MIC for older devices. It will also perform AES, although the method it uses is a little different from the one defined in 802.11i. It has most of the features addressed in the 802.11i section. Some of these include the ability to negotiate a cipher suite or authentication method with robust security network (RSN) information elements.

One of the reasons why 802.11i was not ratified was because of certain requirements that were not well defined at the time. With the WPA standard using whatever the IEEE 802.11i task force had already completed, some changes were needed to be able create this interim standard. These changes led to a number of differences between the two standards.

The first big difference is that WPA supports TKIP by default. This is unlike 802.11i, which supports AES CCMP by default. The next item is the fact that WPA does not even support AES CCMP; it does, however, support AES, just not in the variation that 802.11i calls out. The final major item that differentiates WPA from 802.11i is the RSN IE. This is used to pass the supported cipher settings between the wireless access points and clients. In 802.11i, this portion was not well defined so the WPA standard had to create some newer rules without having them affect anything that might be done to the RSN IE from the 802.11i task group. This was accomplished by creating a WPA IE and using different values to distinguish them from one another. This helped so that once the RSN IE was well defined, it was not difficult to put it into WPA.

## 12.10 802.11i

The recently ratified 802.11i security standard came about in response to a need to improve the security of 802.11 networks to a level sufficient to warrant wireless as a generally accepted secure transport medium. In this standard, the IEEE outlined a secure way to access wireless networks. It also tried to mitigate the now-enormous amount of threats that were making wireless networking a real risk for companies.

Looking at the process of how 802.11i became a standard, one needs to go back to July 1999 when there was interest in enhancing the MAC layer of 802.11 for quality of services (QoS) and privacy. This built up enough steam to create the task group TGe in March 2000. After a year, it was determined that this group needed to split into a security group and a QoS group because of the large workload. This split created the TGi security working group. This group created the 802.11i standard and put it up for approval by voting on it. To proceed to the next level of voting, the working group had to have a 75 percent approval vote. This approval took a number of drafts over a three-year period. Once this was done, there was a level of sponsor ballots and finally the standard board's approval process. On June 24, 2004, the standard body finally approved the 802.11i standard.

Looking at 802.11i up close, one notices that it uses a number of standards, protocols, and ciphers, which have already been defined outside the 802.11i. A number of standards are also defined inside it as well; RADIUS, 802.1x, EAP, AES, RSN, TKIP, and others are some of the defined standards that are part of the 802.11i standard. Some of these are defined inside their own document and some of them are officially created inside the 802.11i document.

The Robust Security Network (RSN) standard is used for dynamic negotiation of authentication and encryption. It is used to negotiate what kind of encryption a client can support as well as what type of encryption is required based on a policy. A deeper look into RSN is available in Section 12.10.1. Before we get into this, we need to address 802.11i standard as a whole.

Another piece of the 802.11i standard is the ability to use EAP. It was determined that the 802.11i standard would not specify an authentication method or type; rather, it would allow a protocol that can perform multiple types of authentication inside itself. This is exactly what EAP does; it allows the use of many different authentication types from passwords, smart cards, certificates, and many others based on the same request, accept, and reject methods. For EAP to work correctly with the 802.11i standard, another well-known standard must facilitate the transmission of EAP between untrusted and trusted entities. This is where the 802.1x standard fits in; its main goal is to provide a framework for strong authentication and key management. The 802.1x protocol allows the access point only to permit an EAP request into the network. This is the case until the client is properly authenticated. Once authenticated, key nego-tiation and subsequently network access can be achieved.

As included in WPA, the 802.11i standard needed an option for envi-ronments in which an authentication server was not financially feasible. This authentication server was a requirement of the 802.1x standard. To make 802.11i viable for both large enterprises and small office/home office (SOHO) users, another method was necessary. This is where the preshared key method originated. This is very similar to WPA and its preshared key method. When a preshared key is used, each client uses a secret to create subsequence-keying material. This master key is the same across the network, just like WEP, although it is used to create a session-based key for each client.

Before getting into the details of how 802.11i works, one needs to understand all the components that make up 802.11i so one is comfortable with a full system overview. Below is each main portion of the 802.11i standard. As said above, most of the standards and protocols inside 802.11i are either their own standard or are located inside the 802.11i standard. The subsections below outline the standards that are located inside the 802.11i standard.

## 12.10.1 Robust Secure Network (RSN)

Robust Secure Network (RSN) was created as part of the 802.11i security standard. RSN specifies user authentication through IEEE 802.1x and data encryption through the Temporal Key Integrity Protocol (TKIP) or Counter Mode with CBC-MAC Protocol (CCMP). RSN also has the option to use TSN to allow for the use of older security methods such as WEP. TSN is explained in detail in Section 12.10.1.1. RSN uses TKIP and AES as encryption methods to protect the confidentiality of data. The TKIP solu-tion is used for backwards compatibility for legacy devices, and the AES is what RSN is using as a long-term encryption method. AES is set up in Counter Mode with the CCMP. AES can be set up and used in multiple ways, so 802.11i stated that AES must be used in a method called CCMP. The RSN protocol also uses EAPOL-Key messages for key management. The description below reveals how RSN works with 802.11i in aiding to choose an available authentication method and encryption cipher scheme.

Advertising the cipher suites supported on an access point and client is done through Robust Secure Network Association (RNSA) messages. These messages spell out the supported ciphers of each party and negotiate what method will be used to connect securely. These messages are located inside what is called an RSN IE or an RSN information element. A RSN IE is used to tell the other devices about what cipher suites the sending

| Element ID | Length | Version | Group cipher suite | Pairwise cipher suite | Pairwise cipher suite list | AKM suite count | AKM suite list | RSN capabilities | PMKID count | PMKID list |
|---|---|---|---|---|---|---|---|---|---|---|
| 1 octet | 1 octet | 2 octets | 4 octets | 2 octets | 4-N octets | 2 octets | 4-N octets | 2 octets | 2 octets | 16 octets |

**Figure 12.15    RSN-IE frame detail.**

device supports. The RSN IE can be sent in a beacon from an access point or in an association request from a client. After an association request, a response will be returned with an RSN IE listing what requesting method matched the method supported by the other party.

The standard allows RSN IE optionally to be inside each of the following management frame types:

- Beacon
- Association Request
- Reassociation Request
- Probe Response

Looking at Figure 12.15, one can see the many parts that make up an RSN IE frame. Of the 11 sections of this frame, only the first three are required in all RSN IE transmissions. After the three required fields, all other fields must have the preceding field inserted with data or the frame will not be properly received. This means that if one needs the value in the ninth field, one must include all eight preceding fields.

- *Element ID.* This field supports 48 decimals or 30 hexadecimal digits. Currently, the only allocated element ID is 48, which stands for RSN.
- *Length.* The Length field identifies the total length of the RSN IE frame. Currently, the frame is only 255 octets long. When using a large number of cipher suites, one may run into a case where one can only support a limited number of cipher suites. This is due to the limit in the total size of the RSN IE frame.
- *Version.* The Version field is used to show what version of RSN is currently being used. It holds up to two octets. Today, only a single version of RSN exists. This version is shown as 1 in this field; Versions 0 and 2 are reserved for other versions.
- *Group Cipher Suite.* This is the cipher suite used to protect broadcast and multicast traffic. It holds up to four octets of information about the total group cipher suites used: up to two octets of information about the multicast and two octets for broadcast cipher suites used.
- *Pairwise Cipher Suite Count.* This field lists the number of selected pairwise cipher suites. It is only two octets long.

- *Pairwise Cipher Suite List.* This field contains all the ciphers that were selected for the pairwise key. Each of the cipher suites is accounted for in the Pairwise Cipher Suite Count field. Each one has a corresponding type inside this field. Each one of the used cipher suites is four octets long. Of the four octets, three are used for the Organizationally Unique Identifier (OUI) field and a single octet is left to identify the cipher suite.
- *AKM Suite Count.* The Authentication and Key Management (AKM) Suite Count is used to determine how many different key management options are available, such as preshared keys or ones dynamically allocated with 802.1x. To note is that in an Independent Basic Service Set (IBSS), only a single AKM suite can exist. This field has a maximum size of two octets.
- *AKM Suite List.* The Authentication and Key Management Suite List is used to specify what key management options are available. Depending on the AKM Count Suite field, this field could have multiple, four-octet sections defining each key management option. Currently, only two options exist: pre-shared keys and 802.1X key management. This leaves a number of reserved and vendor-specific options to include later.
- *RSN Capabilities.* This is a two-octet field used to identify what RSN capabilities are available on the network. It identifies whether or not the device is capable of a pairwise key. It also allows the device receiving the RSN IE to understand if it can or cannot support RSN altogether; if it cannot support RSN, TSN can be tried. In the event that it is not understood, it is assumed that it cannot support RSN.
- *PMKID Count.* The Pairwise Master Key Identifier Count field is a two-octet field used only with reassociation. It is used to cache keys so that when a client is roaming, it does not have to go through the entire authentication process with each access point. This is to speed up the timing and lower the bandwidth as a client roams from one access point to another. The count is to define how many of these credentials are currently inside the RSN IE frame.
- *PMKID List.* The Pairwise Master Key Identifier List field is four octets for every PMKID identified with the PMKID Count field. This is where each of the different types of PMKID is stored. Currently, there are three main PMKID types listed. The first one is a cached PMK that has been obtained through pre-authentication with another AP. The second one is a cached PMK from an EAP authentication. And the third is a cached PMK from a PSK.

### Table 12.3   RSN Cipher Suite Frame Detail

| OUI | Suite Type | |
|---|---|---|
| 3 octets | 1 octet | |

| OUI | Suite Type | Meaning |
|---|---|---|
| 00-0F-AC | 0 | Use group cipher |
| 00-0F-AC | 1 | WEP-40 |
| 00-0F-AC | 2 | TKIP |
| 00-0F-AC | 3 | Reserved |
| 00-0F-AC | 4 | CCMP |
| 00-0F-AC | 5 | WEP-104 |
| 00-0F-AC | 6-255 | Reserved |
| Vendor OUI | Other | Vendor-specific |
| Other | Any | Reserved |

Now let us look more closely at the different types of cipher suites. In RSN, one is only concerned about telling the other party what cipher suites are supported. Once that has been accomplished, one can decide if it is possible to negotiate a common cipher suite or method between both parties. In the RSN Cipher Suite frame section, there are six specified cipher suites and a number of reserved and vendor-specific cipher suites that can be used in the future. Today, the six supported cipher suites are identified with the hexadecimal code 00:0F:AC. Looking at Table 12.3, one can see that each of the following cipher suites is listed as 00:0F:AC and numbers 1 through 6.

The RSN standard is a method to negotiate what types of security methods each client and each access point supports. These security methods are identified as cipher suites inside the RSN IE frame. These cipher suites allow for the use or non-use of any combination of security methods. This means that a policy could be put into place that negates the use of weaker security methods such as WEP and allows for a choice of TKIP or AES. This gives the architects or designers the ability to create a policy allowing or denying whatever cipher suites they might feel are weak or not needed.

### 12.10.1.1 Transition Secure Network (TSN)

The Transition Secure Network (TSN) is part of the RSN portion of 802.11i. It is used to achieve backwards compatibility with older wireless systems.

It was carved out of RSN to provide this backwards compatibility. With RSN, as stated above, it is possible to have a number of authentication and encryption types running on an access point. To make sure that some of the weaker authentication and encryption types were not set up in RSN, they were taken out and considered TSN. This makes RSN more secure and allows an easy way to turn off all the older weak methods. With RSN, if one chooses not to support TSN, WEP will not be included as an option to negotiate between the access point and the wireless client.

## 12.10.2 *Temporal Key Integrity Protocol (TKIP)*

The Temporal key Integrity Protocol (TKIP) was an interim solution developed to fix the key reuse problem of WEP. It later became part of the 802.11i and subsequently part of WPA standards. This meant there were various flavors of TKIP until 802.11i was finalized. One of the first notations about the theory and concepts of TKIP was published in December 20, 2001, by Russ Housley and Doug Whiting, in an article entitled "Temporal Key Hash." This article described the general principle of TKIP, although it was not enough on which to base a standard. That is where 802.11i came in with a more in-depth creation of TKIP.

TKIP was included in the 802.11i standards for backwards compatibility. The 802.11i standard did not want to use a cipher based RC4, so they chose AES (see Section 12.10.3). TKIP was put into 802.11i for the sole reason of helping older devices transition to 802.11i. To do this, 802.11i needed to support a protocol that could easily upgrade WEP to something safe enough to include in 802.11i. One of the main reasons for using TKIP over WEP came from the increased security and increasing number of attacks that were plaguing the WEP protocol. Using TKIP protected against these attacks and reduced the overall risk of operating a wireless network.

The TKIP standard also saw value in the industry because the migration from WEP to TKIP was an easy one. In most cases, moving from WEP to TKIP involved a small firmware change. This meant that no hardware was required to make the change and also that most older, already purchased equipment would be able to upgrade to TKIP.

Another interesting note about TKIP comes from Cisco Systems. Cisco came up with a TKIP solution well before the 802.11i standard defined one. This has led some people to wonder about which version of TKIP is on a certain product. Vendors other than Cisco also created TKIP-based solutions before the standard was ratified. Today, Cisco differentiates its versions of TKIP and the standard one by calling it the Cisco Key Integrity Protocol (CKIP). In Cisco products, one can specify to use TKIP, which is the 802.11i-compliant version, or CKIP, which is the Cisco-created version.

The TKIP encryption portion works in a two-phase process. The first phase generates a session key from a temporal key, TKIP sequence counter (TSC), and the transmitter's MAC address. The temporal key is made up of a 128-bit value similar to the base WEP key value. The TKIP sequence counter (TSC) is made up of the source address (SA), destination address (DA), priority, and the payload or data. Once this phase is completed, a value called the TKIP-mixed transmit address and key (TTAK) is created. This value is used as a session-based WEP key in the second phase.

In the second phase, the TTAK and the IV are used to produce a key that encrypts the data. This is similar to how WEP is processed. In WEP the first 24 bits of the IV are added in front of the WEP key and then used to create an encryption key that is applied to the data. Then the IV is inserted into the packet header. TKIP extended the IV space, allowing for an extended IV field, which holds an additional 24 bits. In the second phase, the first 24 bits are filled with the first 24 bits of the TTAK. The next 24 bits are filled with the unused portion of the TSC. This is safer than WEP because the key is using a different value, depending on who one is talking to. In WEP, each client or access point creates the same random value. Some products never even created a random value and just incremented the value by one, making it an easy target for hackers.

The basis of TKIP came from the WEP protocol. In the 802.11i standard, TKIP is referred to as a cipher suite enhancing the WEP protocol on pre-RSNA hardware. This is espoused because RC4 is still used as a cipher, although the technique in which it is used has improved greatly.

### 12.10.2.1  TKIP Message Integrity Check (MIC)

Similar to TKIP, the Message Integrity Check (MIC) had also many versions before 802.11i defined it as a single standard. Once this was done, MIC became known as Michael although the acronym MIC still remains. Today with 802.11i, ratified MIC is Michael and vice versa. The protocol itself was created to help fight against the many message modification attacks that were prevalent in the WEP protocol. The IEEE 802.11i standard describes the need for MIC in the following quote: "Flaws in the IEEE 802.11 WEP design cause it to fail to meet its goal of protecting data traffic content from casual eavesdroppers. Among the most significant WEP flaws is the lack of a mechanism to defeat message forgeries and other active attacks. To defend against active attacks, TKIP includes a MIC, named Michael." The MIC was created as a more secure method of handling integrity checking compared to the IVC in WEP.

The MIC is a hash that is calculated on a per-packet basis. This means a single MIC hash could span multiple frames and handle fragmentation. The MIC is also on a per-sender, per-receiver basis. This means that any given conversation has a MIC flowing from sender A to receiver B and a separate MIC flowing from sender B to receiver A.

The MIC is based on seed value, destination MAC, source MAC, priority, and payload. Unlike IC, MIC uses a hashing algorithm to stamp the packet, giving an attacker a much smaller chance to modify a packet and have it still pass the MIC. The seed value is similar to the WEP protocol's IV. TKIP and MIC use the same IV space, although they have added an additional four octets to it. This was done to make the threat of using the same IV twice in a short time period less likely.

The MIC is also encrypted inside the data portion, which means it is not obtainable through a hacker's wireless sniffer. To add to this, the TKIP also left the WEP IVC process, which then adds a second, less secure method of integrity checking on the entire frame. To combat message modification attacks, the TKIP and MIC went a step further and introduced the TKIP countermeasures procedures. This is a mechanism designed to protect against modification attacks. It works by having an access point shut down its communications if two MIC failures occur in 60 seconds. In this event, the access point would shut down for 60 seconds. When it comes back up, it would require that all clients trying to reconnect change their keys and undergo a re-keying. Some vendors allow one to define these thresholds, although the MIC standard calls out these values.

To prevent noise from triggering a TKIP countermeasure procedure, the MIC validation process is performed after a number of other validations. The validations performed before the MIC countermeasure validation are the frame check sum (FCS), integrity check sum (ICV), and TKIP sequence counter (TSC). If noise was to interfere with the packet and modify it, one of these other checks would be able to find it first, thus preventing the frame from incrementing the MIC countermeasure counter.

## *12.10.3 Advanced Encryption Standard (AES)*

One can apply AES in many different ways. The way that the 802.11i standard has chosen to apply AES is with CCMP, which is based on CBC-MAC. CCMP was chosen for data integrity and authentication, with the Message Authentication Code (MAC) providing the same functionality as the Message Integrity Check (MIC) used for TKIP. Before diving into CCMP, one needs to look at AES and some of its modes. The first term is CTR; this is AES in Counter mode. This mode is used for confidentiality. The next mode is called CBC-MAC, which stands for Cipher Block Chaining Message Authentication mode. This mode is used for integrity. AES also has combined CTR and CBC-MAC to create CCM. CCM is the acronym for CTR/CBC-MAC mode of AES that incorporates both the confidentially of CTR and the integrity of CBC-MAC.

## *12.10.4 802.11i System Overview*

Having looked at each part that makes up the 802.11i standard, one can now look at it as a whole. One will see how the client connects to the access point, authenticates, and negotiates keys. Each one of these steps leverages the outlined standards discussed thus far.

The client would first need to make a connection to the access point. This would happen through the normal open key authentication process (Section 12.2). Contrary to most 802.11 standards, 802.11i only allows for open system authentication. This is due to the discovery of a security flaw in shared key authentication. This flaw is detailed in Chapter 13.

After the initial connection request, the client would need to hear an RSN IE broadcast or send a probe request with an RSN IE. Whichever way this RSN IE frame is sent, both clients and access points need to negotiate on a cipher suite for use. After sending the RSN IE frames and reaching a negotiation, the EAP process starts. This can start with the access point sending an EAP identity request or a client sending an EAPOL-Start frame. Once the EAP process has started, it will go through the EAP authentication process associated with each particular EAP type. This process is outlined in Section 12.8. It ends with the client receiving an EAP success message from the access point. During this process, an AAA key is sent from the authentication server to the wireless end device. This key is used as a seed key to create the keys outlined below.

The key exchange process takes the original 802.1x EAPOL-Key frame and makes some modifications, allowing for the use of WEP-40, WEP-104, TKIP, and CCMP cipher suites. From the 802.1x section, the EAPOL-Key frame only supports WEP-40 and WEP-104 keys. The 802.11i standard modified this and added the ability for the frame to carry TKIP and CCMP keys as well. A process known as the four-way handshake accomplishes this key exchange. This process takes two main keys and creates unique group and session keys for each client. These session and group keys are created from the two main keys: (1) the pairwise key or the pairwise master key (PMK) and (2) the group key or the group master key (GMK).

In an 802.1x 802.11i setup, the PMK comes from the authentication server. If the 802.11i setup is using preshared keys, then the PMK is mapped to a password. The PMK is divided into three keys. The first key is the EAPOL-key confirmation key (KCK), which is used to provide data origin authenticity. The second key created from the PMK is the EAPOL-key encryption key (KEK), which is used to provide confidentiality. The last key is called the pairwise temporal key (PTK) and this key is also used for data confidentiality. To create the PTK, a pseudorandom function takes place with the access point's MAC address, client MAC address, and a nonce sent from each side as well. This allows a single master key to create multiple session keys without having to re-exchange a new master key each time.

The next key with regard to 802.11i main keys is the group key or group master key (GMK). This key is similar to the PMK except that it is used for beacon and management traffic encryption. The same process of hashing senders' and receivers' MAC addresses and nonces is used to create a group temporal key (GTK) from a group master key.

Having discussed the keys and how they are split up to accommodate session encryption, one can now look more closely at the four-way handshake. This handshake starts with the authenticator sending the supplicant a nonce. This is often referred to as the ANonce in the 802.11i standard. This nonce is a random value used to prevent replay attacks. This means that old nonces cannot be reused. After each party receives a message, the first step before any other is to check and see if the nonce was changed or if the same nonce was incorrectly reused. Once the wireless client receives the first message, it will check the nonce and then generate an SNonce. This nonce will be used in the next step to calculate the pair transient key (PTK). After the PTK is created, the client will then send the SNonce as well as the security parameters outlined in the RSN IE frame to the access point. This information is the second message in the four-way handshake. All of this information will be encrypted using the KCK, which will protect it from any modification while in transit. Once the access point receives this, it will check that the nonce is not an old value. Once this is done, it will also generate the PTK from the SNonce and ANonce, and then check the KCK to make sure it was not modified in transit. Once this is done, the third message in the four-way handshake will take place. This message is used to tell the client to install the PTK key that was created and, if used, this message will send a GTK to the client to install. Once the client receives this, it will check the KCK and, if it is correct, install the key or keys. The last message is a confirmation used to let the authenticator know that the client has successfully installed the keys and is ready to communicate using them. Figure 12.16 shows this four-way handshake as it takes place.

## 12.11  Wi-Fi Protected Access (WPA2)

After 802.11i came out, the Wi-Fi Alliance wanted to continue the initial investment made in WPA. This created an issue because the 802.11i standard was now out and another standard was not what the industry needed. To keep WPA going, the Alliance decided it would go back to the core benefit the organization provided — standard interoperability testing and certification. In creating WPA2, the Wi-Fi Alliance made this

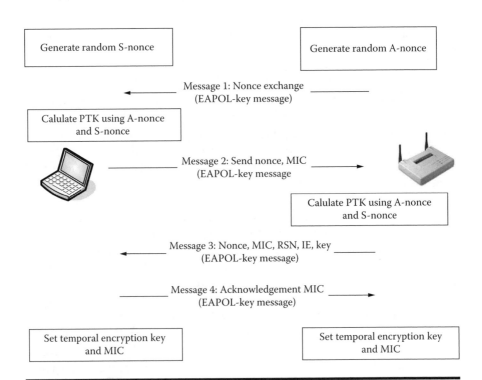

**Figure 12.16   802.11i Four-way handshake.**

version of WPA an interoperability mark similar to Wi-Fi. This mark ensures that any product carrying it has an interoperable 802.11i standard.

## 12.12  WLAN Authentication and Privacy Infrastructure (WAPI)

China has decided that 802.11i is taking too long and they are better off creating their own standard. Even with 802.11i out today, China has still said they want to take their own approach. This has led to the creation of the WLAN Authentication and Privacy Infrastructure (WAPI) standard. This standard has many similarities to 802.11i, such as RADIUS and 802.1x. Getting any information about this standard is a violation of Chinese national security under China's State Council Directive 237, which regulates commercial encryption ciphers and requires encryption technology to be developed and sold under a blanket of secrecy.

To sell WLAN products within China, foreign equipment vendors must license WAPI through an agreement with one of 24 Chinese companies granted rights by the Chinese government. This has been a hard sell to many industry leaders such as Broadcom, Texas Instruments, Intersil, Atheros, and many others. The Chinese government has asked them to provide all the intellectual property involved in their chipset. This intellectual property has made them a market leader; and to sell in China, they could lose that. As one might guess, many of these companies are not taking the risk.

## 12.13 Rogue Access Points Detection

Detecting rogue access points has become almost an industry of it own. Many products came to market with the sole purpose of identifying rogue access points. When other companies saw these moves, they tried to adapt their current products to perform this function as well. Today we see a number of products that perform rogue detection from access points with built-in detection, to appliances, to stand-alone devices whose sole purpose is rogue detection.

In the chapter on risks and threats (Chapter 2), there was very little discussion of rogue access points, merely what they are and how they pose a risk to the enterprise. This section discusses how to detect rogue access points. Each company that developed rogue access point detection software took a unique approach in doing so. This has given us a wide variety of techniques to use in detecting and combating rogue access points.

Some products, such as Cisco's wireless LAN solutions engine (WLSE), have the ability to detect and prevent rogue access points from connecting to the network. How WLSE works is by using other access points, which are managed by WLSE, as sensors. These access points listen for beacons or data frames coming from other access points that are not defined in the WLSE console. When this happens, the unknown access points are labeled as rogue. The next step, which is where Cisco has differentiated itself from other rogue detection systems, is the ability to leverage the wired network to shut down rogue access points. This is done by querying the Cisco switches in an attempt to find the rogue access point's MAC address. This MAC address is easily read from the airwaves so WLSE already knows what it is looking for. When the query is made to the network, the switch, which has the access point's MAC address in its Content Addressable Memory (CAM) table, will respond by saying that "I have that MAC address." WLSE can then tell the switch to shut down that port and thus disconnects the access point's connection to the network. This method can fight a moving rogue access point very well. As soon as someone moves the access point somewhere else, the whole process starts over finding, identifying, and then shutting down the rogue access point.

Some of the other approaches that have been made to go beyond the location and identification of rogue access points have to do with using denial-of-service attacks. This feature is interesting and might be considered illegal. This is definitely something that someone in a multi-tenant location may want to consider before using. This approach works on the basis of finding a rogue access point and then using nearby access points to send disassociation messages to all clients connecting to the rogue access point. As stated, in a multi-tenant location, this could mean disrupting a neighbor's network. This could become a legal issue quickly. So far, no court cases have come out of the woodwork with regard to this particular issue, although it is just a matter of time.

## 12.14 Chapter 12 Review Questions

1. Which encryption mode was selected for 802.11i?
   a. RC4
   b. AES
   c. AES-CCMP
   d. 3DES

2. Which EAP type is part of the EAP RFC?
   a. LEAP
   b. EAP-MD5
   c. EAP-TTLS
   d. EAP-TLS

3. What EAP method did Microsoft, Cisco, and RSA create?
   a. PEAP
   b. LEAP
   c. EAP
   d. EAP-TLS

4. Does China have its own wireless security standard?
   a. Yes
   b. No

5. In 802.1x, what is the RADIUS server often called?
   a. Supplicant
   b. Authenticator
   c. Authentication server
   d. RADIUS server

6. In 802.11i, what mode is one using if WEP is enabled?
   a. RSN
   b. TSN
   c. TKIP
   d. CCMP

7. Which IEEE standards deal with security? (Select two)
   a. 802.11
   b. 802.11b
   c. 802.11i
   d. 802.1x
   e. 802.11g
   f. 802.16

8. During the time frame when WPA was released and 802.11i was still being finalized, what reservations would an IT manager have about using TKIP?
   a. Weak keys
   b. Lack of interoperability
   c. Possibility of repay attacks
   d. Short age

9. Which EAP type is used to create a tunnel in which other, older authentication methods can take place?
   a. PEAP
   b. EAP-MD5
   c. EAP-TTLS
   d. EAP-TLS

10. Which of the following is not a VPN technology term?
    a. IPSec
    b. SSL
    c. ESP
    d. AH

11. Which of the following provides layer 2 security by allowing for unique and changing encryption keys?
    a. 802.1x/EAP
    b. WEP
    c. TKIP
    d. AES

12. During an EAP negotiation, what happens after the identity response packet is sent from the client to the access point?
    a. The RADIUS server sends out a challenge.
    b. The access point sends out a challenge.
    c. The access point forwards the message to the RADIUS server.
    d. The access point converts the message and sends it to the client.

13. What is a new federal encryption standard that is used in a certain mode on wireless networks?
    a. RC4
    b. WEP
    c. DES
    d. AES

14. What is used to ensure the integrity of packets traveling across the airwaves?
    a. Authentication server
    b. RADIUS
    c. MIC
    d. IV

15. WEP keys can also be thought of as _____.
    a. Tokens
    b. Secure IDs
    c. Key cards
    d. Shared secrets

16. Most wireless equipment is set to which better authentication method?
    a. Open key authentication
    b. Shared key authentication
    c. Dual key authentication
    d. WEP key authentication

17. Which of the following can be used with the 802.1x standard for wireless authentication?
    a. TACACS
    b. RADIUS
    c. Kerberos
    d. SSL

18. A WEP key is _____.
    a. Strong
    b. Secure
    c. Weak
    d. Safe

19. What is EAP-FAST replacing?
    a. PEAP
    b. LEAP
    c. WEP
    d. EAP-TLS

20. What IEEE group is responsible for creating WLAN security standards?
    a. 802.11a
    b. 802.11g
    c. 802.11ii
    d. 802.11i

21. Which of the following is not a correct EAP type?
    a. EAP-MD5
    b. PEAP
    c. LEAP
    d. EAP-TTTLS

22. What security solution provides for unique changing keys that change to prevent an attacker from cracking the key?
    a. IPSec
    b. 802.1x
    c. Dynamic WEP
    d. RADIUS

23. What wireless security cipher is approved by the government for standard use?
    a. WEP
    b. RC4
    c. AES
    d. TKIP

24. What IEEE standard provides the authentication framework for all IEEE 802-based networks?
    a. 802.11i
    b. 802.1x
    c. 802.5
    d. 802.11
    e. 802.11b

25. What EAP type creates an encrypted tunnel and then performs another EAP exchange inside this tunnel?
    a. PEAP
    b. EAP-TLS
    c. EAP-TTLS
    d. EAP-MD5
    e. LEAP
    f. EAP-FAST

# Chapter 13

# Breaking Wireless Security

This chapter looks at breaking wireless networking through the eyes of a hacker. It discusses most of the flaws that have been found relating to the many different wireless security methods and the security methods identified in the previous chapter to get a full understanding of how they can be defeated. Having an understanding of what is currently possible and what a hacker can do will help in evaluating the amount of risk surrounding any particular wireless network.

## 13.1 The Hacking Process

This chapter defines the process taken by an attacker, a process that has been long-standing in the hacker community. With wireless networking providing a new way into networks, some of the structure of this process has changed. No longer is a network target examined and attacked. Today's wireless war driving has shown that attackers will now find a network first, then plan an attack, and finally return to attempt a break-in. The process outlined in this chapter follows the standard hacking process, with the clear thought that a wireless network will be used as an avenue of entrance.

## 13.1.1 Information Gathering

Step one in understanding wireless hackers has to do with how they would find a target as well as the information online about the target. This step is called information gathering. Most likely, a wireless hacker will find a target by war driving. In this scenario, an attacker is neither looking for something specific nor is he targeting a company directly. The attacker just happened to drive by and wants to either test out the security or is just looking for a free Internet connection. Chapter 2 discussed why war drivers do what they do and how they go about doing it. In this chapter, we expound on this by looking at the next steps that occur after detecting a wireless network. When an attacker war drives a network without planning it, this actually gives the company the upper hand. This is because the attacker's level of knowledge about the company is very limited. This upper hand will not last long if the attacker decides to plan an attack on this network.

After an attacker has decided to plan an attack, information gathering is the next step. To gather information, an attacker would use the largest amount of data ever put together, the Internet. Most people are unaware of the Internet's ability to find information about companies or people themselves. Have you ever "googled" yourself? Try it one day with your first name and last name in quotes (e.g., "Aaron Earle"). What comes back might amaze you. An attacker will do this with a company's name and, lo and behold, he will find a slew of information about that company. What the attacker is really looking for is IT people from the company trying to get help about a product. If someone at this company needed some help with a product, one of the steps might be to try a message board for the answer. Most likely, they will leave an e-mail address or, in some cases, the e-mail system may insert a disclosure message automatically before the e-mail leaves the system. This message most likely has the company name inside it, as well as a legal disclaimer.

If an attacker is able to find this information, he can assess what types of networking products a company is using. In this case, the most relevant information would be e-mails to message boards about the company's wireless network. The attacker can continue to gather useful information using other avenues of information gathering, such as public records. Going to the Securities & Exchange Commission's Web site would be an example of public records. This site contains specifics on recent mergers and acquisitions, information that could serve as another avenue of attack. Also state government Web sites. Most likely, one can find information about a company and its employees as well as the names of management personnel. If two companies are linked together through a search engine, this might indicate that they are close partners. If this is true, there might also be a network connection between them. If a hacker found that the partner of the target has a weak network security system in place, he might use it as a backdoor into the target's network.

Another type of information-gathering technique comes from domain registration information. When companies register their domain names, they must include information about themselves. This information is publicly available and can serve as a means of social engineering. There are a number of tools to perform this type of information-gathering attack. The most common are whois, nslookup, dig, etc. A freely available tool called Sam Spade can perform all these look-ups, and more. Sam Spade is available at SamSpade.org. Web site. Once downloaded and installed, it can be run to perform any of the information-gathering techniques listed above to find domain registration information.

Now take a look at Dun and Bradstreet (D&B), a company that creates, tracks, and monitors business credit. If an attacker is willing to spend around $25, he can pull a small credit report on a company. He could also spend more and get a more encompassing one that details partners, creditors, or any legal trouble the target may be experiencing. For the example here, however, the small report will work. What an attacker can do with this information is find out the business owner's name as well as key senior members who are listed on the credit report. Once an attacker has this information, he can perform social engineering attacks on helpdesk personnel to reset passwords.

## 13.1.2 Enumeration

Enumeration, footprinting, and scanning are all addressed in this section and defined hereafter as just enumeration. After an attacker has finished gathering information about the company he wants to attack, the next step called enumeration, takes place. In enumeration, the attacker goes looking for everything he can connect to and tries to understand what type of products there are, what company made them, and what software or firmware versions they are running. This is the step where war driving comes into play. A war driver is in the enumeration stage when he scan a wireless network. A war driver may have skipped the information-gathering stage altogether and go right into the enumeration stage just by war driving an area and not a selected target. A war driver may often find a network with war driving and then go back to information gathering before actually breaching the network.

Looking into the enumeration stage, there is another interesting step switch or process repeat. In the enumeration stage, one may only be able to enumerate so long until one is required to move to the next step of compromise. If an attacker has exhausted his enumeration capabilities, he may have to compromise a device to gain access into another portion of the network. At that time, the attacker may need to revert to the enumeration stage to test all the new devices that he can connect to now that a certain network device has been compromised. This is very relevant in wireless network hacking. When an attacker can enumerate a wireless network, he can only go so far; after that point, he must compromise the wireless to gain access to other devices on the wired network. Once access has been gained, the enumeration process must start over on the newly discovered wired devices.

There are many tools used in enumeration, most of which are some type of scanner. The different types and functions of scanner are numerous. See Chapter 15 for a discussion of a number of scanners for wireless networks. Scanning wireless networks is a small subset of what a scanner can do. The type of scanner most widely known is the port scanner, which scans for open TCP or UDP ports. After identifying these open ports, an attacker can use another scanner to finger or footprint the device by requesting and examining the information sent from these open ports. Most operating system (OS) communication across these open ports is unique to the OS manufacturer. A scanner with the ability to finger or footprint a network can examine the traffic responses from a device and understand what operating system it is by examining the traffic. This is because certain operating systems handle TCP/IP in different ways and create certain packet formats that give away the identity of the OS.

Another type of scanner is one that uses ICMP and SNMP to map out networks and networking devices. This type of scanner is used to create a network map so the attacker understands the company's network layout. This scanner is sometimes included with a port scanner and needs to run prior to a port scan to know what device to port scan.

The final type of scanner discussed here is the security vulnerability scanner. This type of scanner can test a network device for a large number of known exploits and vulnerabilities. This is usually performed in a very short period of time. This is the last sub-step of enumeration; once this is done, all that is left to do is exploit the attack that the tool identified. This is the vulnerability where scanners stop short; they only identify a vulnerability and will not actually attempt to perform the next operation, called compromise. Some tools have come out that can actually execute these exploits.

### 13.1.3 Compromise

Now the attacker attacks the wireless network or another network device to take it over; once it is taken over, it is considered *compromised*. If an attacker can compromise a wireless access point, he can insert himself inside the network and go after the next network target.

The compromise section has limited tools available to perform the actual compromise. The previous section showed that some scanners could report on the vulnerabilities of a device, although the majority of them do not actually perform the vulnerability, but rather test that the vulnerability can be performed. To compromise a device, one must create or find an exploit. If one finds an exploit, it would likely have to be compiled and executed. To find these exploits, one needs to know where to look; these exploits are available in underground Web sites or IRC channels. Most of the time after an exploit has been out in public view for a considerable amount of time, a tool may appear that is point, click, and compromise.

This stage of the process is where the real meat of this chapter lies. After completing the entire hacking process, one will come back to this section, looking at each of the different types of wireless networks and how they can be compromised individually. Before doing that, one needs to explore the remainder of the hacking process.

## 13.1.4 *Expanding Privileges and Accessibility*

This step is very important in the hacking process. When looking at hacking from a wireless prospective, one needs to understand that to get to this stage, a compromise of the wireless network has already taken place. If a wireless network is compromised, then there is little else to do in relation to it. The attacker would most likely move on the next network target. This stage of the process has more weight in relation to servers and other network devices where access is achieved, although it is not full unrestricted access. The process of expanding privileges deals with attacking a device as a restricted user and taking advantages of flaws to elevate the privilege level to unrestricted.

The next portion of this section deals with accessibility, or being able either to get back into a network or to find an easier way to control a compromised device. When one looks at getting back into a network, one starts to think about backdoors. There are many different ways to set up backdoors. Looking at wireless backdoors, an attacker could set up multiple backdoors in multiple places. In allowing easy accessibility to compromised devices, one needs to look at how this can be achieved. Using tools such as VNC and NCAT on devices can allow an attacker to receive remote shells. Once an attacker gets a remote shell, he can install or activate some type of remote management software. This software will allow for remote GUI OS manipulation. Looking at how to go about installing this is outside of the scope of this book; just be aware that it is part of the process and needs to be done in order to place some types of backdoors.

Now take a look at another interesting backdoor example. This has to do with dual homed wired and wireless computers, such as laptops. If an attacker can find his way into a company laptop, he might be able to find one with a network connection and a wireless connection. With almost all of today's laptops shipping with wireless network cards, the odds of finding this is quite high and very likely. If an attacker can find this and compromises it, he can set up the wireless computer as ad hoc and set up routing into the network. Now when the attacker connects back into the network, no matter what level of wireless security is present, an attacker can circumvent it. To get access to the laptop originally, the attacker may decide to attempt to hack into it at a hotspot or steal it temporarily and return it before anyone finds out. The most likely situation would be where the attacker hacks into a network through some misconfiguration and then attacks the laptop in case the misconfiguration was detected and fixed.

The next backdoor is one of the most common and, luckily for the attacker, in most cases this backdoor already exists. This backdoor is rogue access points; they are not only used by employees who want wireless, but also by attackers who stumble across them. Most of the time, these access point are deployed right out of the box with no security and default settings. This makes the network an especially easy target for an attacker. Preventing and tracking down rogue access points is addressed elsewhere in this book. The main reason that rogue access poses a threat and subsequent risk is that an attacker can use them as easy, already set-up backdoors into a corporate network.

There is one other type of backdoor that is very similar to the rogue access points already discussed; expect that this one is placed by the attacker. If an attacker can gain access into the physical building, he can place a rogue access point. This is common in public buildings such as schools, hospitals, and other public-welcoming businesses. In this scenario, an attacker would purchase an inexpensive access point and find a spot to hide it. Then the attacker could mount his attack from some distance away from the physical security the building might have. With a high-gain antenna connected to a laptop, an attacker could be physically located far away from any physical security force's line of sight.

### 13.1.5 Cleaning up the Trails

This is the last section in the hacking process; it is where one learns what the hackers do to clean up after themselves. If companies were to catch on to the existence of an attacker, it would be through log files and advanced security tools such as IDS and IPS systems. In this section, the attacker would purge logs and perform other housekeeping techniques to hide their existence. This step is required in the hacking process to prevent a company from figuring out that an attacker has broken in. If this step is not performed, a large amount of evidence would remain as to how the attacker got in, where they went, and what they accessed.

When we explore this section in relation to wireless, most of the same ideas and techniques apply. A wireless attacker would try to cover his tracks by changing the access point's log file. This would have to extend to other network devices as well. The switch port would hold the attacker's MAC although he entered through the access point. Any servers the hacker attempted to access would have log files indicating such an attempt. All of these devices would need to have their logs cleaned up. To make things even trickier is the existence of a syslog server. If a syslog server was present, all devices would send their logs to it. This means the attacker would have to attack the syslog server as well if he wanted to clean all trails of his intrusion.

One of the misconceptions of the track covering section is how most hackers are found out. What some attackers tend not to realize is that if they attack a box and are unable to compromise it, their attack will be logged. That log will remain on the device for some time. Some of the ways that the log would be removed include:

- Someone goes through log file and manually erases it.
- The attacker manages to compromise the device and change the log.
- The log has a specific setting that allows it to be overwritten after it has taken up a specific amount of disk space.

An attacker has many tools developed over the years to remove specific entries or repopulate log files with bogus entries. These tools are commonly used to cover the tracks of hackers once they have compromised the device. There are different methods as to how they work. Some tools actually change log events; they would change the event ID or some other part of the log record to make it look like it belongs. Another method involves erasing the entries themselves; some tools allow attackers to erase certain log records that deal with their access attempts. The final kind of log-cleaning tool discussed here performs this function in an obscure way. It populates the log with a large number of bogus entries to make searching for the attacker entries a long and drawn-out process.

## 13.2 Wireless Network Compromising Techniques

This chapter section explores specific techniques used to compromise wireless networks. It details wireless security methods, looking at what vulnerabilities, flaws, and exploits are out in the wild and used to compromise the different wireless security methods. With this knowledge, a proper risk assessment can be made against the many different security protocols, methods, and standards. Compromising a wireless network is only an avenue into the network. When an attacker attacks a wireless network and successfully gets access, one must understand how he got in and how to prevent his access in the future.

The section will address each of the different security flaws in most of the recent wireless security methods, protocols, and standards. This information will show in detail how the exploit, problem, or attack works; how hackers use them; and what can be done to prevent them. Most of these attacks have the only secure work around of using a different security method all too together.

### 13.2.1 WEP

WEP has undergone major scrutiny and subsequently failed all three tenets of its original design. This is observed in each of the areas of confidentiality, availability, and integrity and outlined in this section. Under confidentiality, we explore the known plaintext attack, shared key authentication attack, double encryption attack, man-in-the-middle attack, and a dictionary attack. When looking at availability, we discuss a number of denial-of-service attacks that are mounted against wireless local area networks. Finally, when exploring integrity, we look at some message modification attacks.

#### 13.2.1.1 Stream Cipher Attack

The first attack against the confidentiality of wireless WEP is the stream cipher attack. This attack is based on the work and subsequent article published by Fluhrer, Mantin, and Shamir (FSM). This attack is also the basis of the WEPCrack, Airsnort, BSD-AIR Tools, and many other WEP cracking tools. The attack works in a passive manner in which no one can detect that it is even taking place. This means that trying to find out if this attack is running is nearly impossible. This works because the device running the attack does not need to transmit; instead, it only needs to receive network traffic.

The attack works by taking advantage of a number of weak IV numbers that are used in the general sequence of WEP. These IVs are used as a mechanism to create a different key for each packet transmitted. There are about 9000 interesting IV numbers out of the available 16,777,216. The reasoning behind why these 9000 IV numbers are considered interesting is the fact that the IV has "FF" in the middle of the IV sequence. This IV sequence is defined as three groups of two hexadecimal digits separated by a colon. An example would look similar to the following: A3:4D:33. An interesting IV would have FF in middle and look like this: 3A:FF:5E. When one looks at any IP packet, one can see that RFC 1042 requires an 802.2 SNAP header. This header is 0x88; because one knows what the plaintext is as well as the encrypted portion of it, one can get information on the first key bit. How this works is that the FF comes out to all one digits in binary and when one takes the header portion, which in the SNAP header case is inserted into the data, one gets a piece of the frame in which one knows what the first couple of bytes are before and after they are encrypted. With this knowledge, one can perform what is called a cleartext cryptanalyst attack. The topic of cleartext cryptanalyst is addressed in more detail in the next section. Once that information is known, the process to get the next key bits becomes a guessing game; however, this guessing game has a high probability of correct guessing. At first, the guessing may only yield a 5 percent chance of a correct guess, although as more key bytes are exposed, the percentage of success rapidly increases. A more detailed explanation of the mathematics of how this is achieved is available in the article entitled "Weaknesses in the Key Scheduling Algorithm of RC4."

### 13.2.1.2 Known Plaintext Attack

The known plaintext attack is similar to what the FSM attack uses as a starting point. This attack is possible when one knows or has access to the cleartext and the encrypted text of an information exchange. Having both the encrypted and unencrypted information allows one to perform this attack and subsequently derive a key.

When the text is originally encrypted, it performs an XOR operation that mixes the key with the data. This process is reversible from encrypted text to cleartext just as it is originally applied from cleartext to encrypted text. What this means is that the same process used to encrypt can also decrypt. When the text is broken down to binary, one can perform some binary math and compute the key that was used. Looking at Table 13.1, one can see that the plaintext and key stream are broken down to binary, and binary math is performed to produce the ciphertext. If one were to place the ciphertext below the plaintext and perform the same operation, one would get the key stream. This is how the plaintext cryptanalysis attack works.

To see this attack in action, there are a couple of scenarios in which it might take place. All of the scenarios of this attack fulfill the two main requirements needed: one plaintext packet and its encrypted counterpart.

**Table 13.1  Plaintext Cryptanalysis Detail**

| Known Plaintext Attack Example: |
| --- |
| ▪ 1110100010101010111010 <--- plaintext XOR<br>▪ 1001010101010101010001 <--- key stream EQUALS<br>▪ 0111110111111111101011 <--- ciphertext |
| Now take the plaintext and ciphertext and add them in binary to get the key: |
| ▪ 1110100010101010111010 <--- plaintext XOR<br>▪ 0111110111111111101011 <--- ciphertext<br>▪ 1001010101010101010001 <--- key stream |

Once one has both packets, one can perform the operation detailed above and derive a key.

This first and most predominant scenario is the shared key authentication attack. As learned in the section on authentication, shared key authentication sends a cleartext packet and waits for its encrypted counterpart so it can verify it was encrypted with the correct key. Other methods of this attack are also achievable by looking for predictable IP traffic. For example, when a machine first boots up, it will attempt to log in to the network; this log-on sequence is known text. The first frames that are sent before the user or machine authenticates is the request to authenticate. There are many types of known traffic, such as DNS, Logon, DCHP, etc. All an attacker needs to do is understand when this type of traffic takes place. They can do this by assuming that a given number of frames will take place at certain times. Another option is that this attacker can be run on a number of frames; and when one looks like a normal non-encrypted packet, the process worked.

One of the last ways we are going to look at to perform this attack is to generate traffic. This is done by either sending ICMP, telnet, or any other known protocol. These attempts are only available to one who is already on the wired portion of the network. To get around that, one needs to look at another form of traffic — e-mail. To perform this method, one needs to know who a certain MAC address is and what their e-mail address is. This information can be sniffed out of the air from a hotspot or procured in a number of other ways. Once it is known, just send an e-mail and perform the attack on all traffic until the frame that contains some of the e-mail is captured and the attack succeeds. This could also be done by performing a companywide spam, although this will most likely be caught in a spam filter.

An example of how the shared key authentication attack works is outlined as follows:

1. Alice tries connecting to the network.
2. The access point sends out a cleartext challenge.

3. Alice takes the challenge packet, encrypts it with her WEP key, and sends it back to the access point.
4. Evil Bob extracts the IV (sent in the clear) and key by XORing the challenge with Alice's response.
5. Now evil Bob tries connecting to the network.
6. The access point sends out a challenge string.
7. Now that evil Bob has derived the key from the plaintext cryptanalysis, he can correctly respond to the access point's challenge.
8. Bob connects to the network.

### 13.2.1.3 Dictionary Building Attack

Another attack on WEP is the dictionary building attack. This is performed using the plaintext cryptanalysis attack discussed previously. Once this has been done, any frame with the same IV as the one previously cracked can also be cracked with the same key. This means one can perform the plaintext cryptanalysis attack on multiple frames until one has cracked everything in the IV space. This is not as big or as time consuming as one might think; it can be done with a 24-GB database. Once this has been accomplished, all the frames on the network are now seen by the attacker as cleartext. This will remain true until the WEP seed key is changed on all clients and access points.

### 13.2.1.4 Double Encryption Attack

The double encryption attack takes advantage of the fact that the same key is used to both encrypt and decrypt. To perform this attack, a frame must be captured out of the air that is considered important. Most likely, this frame will be something of value because one can only do this attack one frame at a time. After the frame has been identified, one must change the header to have a destination MAC address that is another wireless client. After this, the attacker must wait for the IV to reset to one minus the original IV. At this point, the attacker can replay the captured frame onto the airwaves. When the access point sees the frame with the expected IV, it will perform the encryption process, actually decrypting the frame instead of encrypting it. After the access point has performed the encryption process, it will forward the now cleartext frame across the air to the forged MAC address specified by the attacker.

### 13.2.1.5 Message Modification Attack

In the WEP process, the Integrity Check field (IC) is used to verify the message's integrity. This 4-byte value can tell the access point if the frame

238 ■ *Wireless Security Handbook*

had corruption or not. By default, an access point will drop a frame with a wrong IC without any logging. This is because of the large number of incorrect transmissions associated with wireless communications. Another issue with the IC is that it is independent of the master WEP key and the IV. Because of this independence, modification can easily take place.

To perform a message modification attack, the attacker must first capture an encrypted packet destined for a different subnet. This is so a router will have to examine the packet. After capturing an encrypted packet, the attacker must modify a single bit and attempt to resend it. Most likely, the modification will offset the IC and the packet will be dropped. After attempting a number of times, the bits that are flipped will make the IC correct again, although the packet itself will be unreadable. The amazing thing about this is that the attacker can try many times without any logging or notification on behalf of the access point. Once the packet passes the access point's IC check, it will go to the router; the router will see that the packet is malformed and send a predictable response back to the original sender. When this response comes across the airwaves, one will have the cleartext and associated encrypted text packet. This will give one what is required to perform cleartext cryptanalysis.

## 13.2.2  Denial-of-Service (DoS) Attacks

Now we will take a look at denial-of-service (DoS) attacks against wireless. Part of all 802.11 networks is the existence of management frames. These frames tell clients that they can connect or must disconnect. As learned in the management frame section, the de-authentication frame will disassociate a wireless end device from an access point. These frames, like all management frames, are in cleartext even when WEP is applied. This means they can easily be faked to force legitimate users off the network. This can be accomplished in a number of ways. The first way to look at it is to replay a previous disassociation frame with a wireless sniffer. An easier method is to use a tool called WLAN-JACK, which can do this for you. Either way, all that is required is the ability to send a de-authentication frame to a wireless client.

On May 13, 2004, a major attack on 802.11/b/G and mixed mode G was found. The flaw affects the clear channel assessment (CCA) procedure, which minimizes the probability that two wireless devices will broadcast on the same frequency at the same time. This attack can cause all devices in range to stop working until the attacker stops releasing the malicious frame. Any device is capable of this attack; and even with most of the current security standards out today, no 802.11, 802.11b, or 802.11g in mixed mode network is safe. The way to prevent this risk is to use a system that performs encryption on all traffic at layer two.

### 13.2.2.1 EAP DoS Attacks

Some of the DoS attacks existing today take advantage of weaknesses in the EAP architecture. These involve sending corrupt frames or engaging the access points in unnecessary processing. Following are some of the types of attacks that can be performed.

The first attack involves sending EAP Stat frames to an access point. If the access point cannot properly process all these frames, there is the chance that it might reboot or become inoperable. This is when an attacker would try to flood traffic past the device as it reboots or malfunctions. Another attack against the access point involves sending malformed EAP messages. Some types of malformed EAP messages can take down an access point or a RADIUS server. This has been proven on Free RADIUS when one sends an EAP-TLS frame with the flags set in a certain way. One of the latest attacks against the access point involves filling up the EAP identifier space. EAP allows 255 ID tags to keep track of each client instance. If an attacker can flood the access point with a large number of connections filling this counter, a DOS attack may be possible.

## 13.2.3 MAC Filtering Attack

After the initial weaknesses of WEP were in the media and out in front of all the IT people, some of them started using MAC filtering as an interim solution until a standardized one was created. This filtering was not without its own problems. A MAC address can be easily changed. Some forms of Microsoft operating systems allow the registry to make these changes. UNIX variants also have an operating system setting to change MAC addresses. On top of both of these known methods to change MAC addresses, many point-and-click tools exist to perform MAC address changing.

All MAC addresses are also easily seen with a network sniffer. This means if someone was to employ MAC filtering in their wireless network, it could be broken as follows. The attacker would use a wireless sniffer to find what MAC addresses are talking on the network. Once this was determined, they could easily change their MAC address by any of the means above or with the help of an automated tool. One tool for this is outlined in Chapter 16. Once they have changed their MAC address they can send a De-authentication frame to the original user or just access the network with conflicting MAC addresses. Either way, they have already circumvented the existing security.

## 13.2.4 Cisco LEAP Vulnerabilities

Cisco LEAP has a number of discovered vulnerabilities. One of them was so bad that Cisco systems gave up using LEAP and went to a new EAP type called EAP-FAST. LEAP was developed to allow for fast roaming from cell to cell.

As discussed in the wireless security history section, a number of tools emerged that defeated the security of LEAP. The first one released was called anwarp. This tool showed that LEAP had no mechanism to prevent an attacker from trying a large number of authentication attempts against the access point. One thing that helps mitigate this threat is that many RADIUS servers have the ability to prevent this. They do so by setting an invalid authentication attempt attribute and a subsequent lockout duration after achieving the invalid authentication threshold.

The next tool released was called asleap; this tool took the above attack further and allowed for an offline attack to take place against LEAP frames. The LEAP vulnerability that this tool took advantage of was a modified version of MS-CHAPV2 that Cisco used to authenticate users on a LEAP network. It was built upon the fact that Cisco used padding for 21 bytes; this makes performing an attack on a hash much easier because the last seven bytes only have two charter options. Because of this, the time it takes to perform a dictionary attack on LEAP is significantly reduced. This makes tools, like asleap, that performed this exploit extremely fast.

## 13.2.5 RADIUS Vulnerabilities

RADIUS uses a shared secret to communicate between a RADIUS server and a RADIUS device or client. This shared secret is created with a hashing algorithm called MD5. The shared secret is calculated from the code, ID, length, request authenticator, and attributes for the response authenticator. All of these values are inserted into an MD5 hash, producing an encrypted string. Because an MD5 hash is a one-way hash, one cannot break it, although it is possible to take a brute-force tool and run every combination of passwords through an MD5 hash and compare them to the original hash. If they match, then the password guessed was the correct one. This holds true on a RADIUS frame because all the values listed above can be captured with a wireless sniffer with the exception of the password, which is inside the hash.

If an attacker can capture the authentication process, he can take the frames offline and perform a brute-force attack on them. This frame, by default, is not authenticated by the RADIUS server. This means anyone can start the RADIUS authentication process and capture the hashed shared secret. One note is to have a correct IP address defined in the RADIUS server; this is a trivial spoofing activity that even a meager attacker can accomplish. The industry has released RFC 2869 to help with this vulnerability. This RFC allows one to set the Message-Authenticator attribute. With the actions of this RFC implemented, the Access-Request packets are only accepted from a valid predefined RADIUS client. This prevents the RADIUS server from responding and giving away its shared secret to anyone who asks for it. Also, once the shared secret is known for a RADIUS conversation, a similar MD5 function can be accomplished on the user's password.

Another vulnerability of RADIUS has to do with its usage as an authentication server for other media types. Most commonly, RADIUS is used for some type of remote access. In the scenario where RADIUS is used for remote access, some administrators have used the same shared secret for both purposes. If the remote access protocol has vulnerabilities associated to it and an attacker is able to compromise the shared secret, then he can use that to connect a rogue access point to the network and capture a user's authentication attempts.

Another RADIUS issue that affects security is the security of the RADIUS server itself. Often, the RADIUS server is set up and maintained by network engineers and not server administrators. Network engineers frequently do not have the skill set needed to secure a RADIUS server. This often happens in big companies where the network engineers and server administrators do not interact with each other on a daily basis. This frequently leaves the security server vulnerable to a number of server exploits.

## *13.2.6 802.1x Vulnerabilities*

The 802.1x protocol had experienced a number of vulnerabilities when the industry decided to adapt it to wireless. The protocol was originally used for wired port-based authentication, not wireless. These vulnerabilities came about from the initial thought that all network equipment would be physically secure, locked into a rack inside a telecommunication closet or data center. All the user would see and connect to was a wall jack. That wall jack could most likely be considered secure, and a general understanding that it is connected to a company switch and only a company switch. In the wireless world, this determination cannot be made because the access points are placed throughout a facility and clients can connect from wherever there is an access point sending signal. This means the existence of rogue access points set up by attacks is a real threat. Some way to validate that the access points are those belonging to the company was needed. When the 802.1x protocol was created, this thought process was never taken into consideration, and a validation step to ensure that users connected to the correct piece of networking equipment was never created in the standard. As of today, the 802.1x standard along with the EAP standard have made a number of modifications to better accommodate wireless communications.

Because of the issue of allowing any piece of network equipment to communicate with an authentication server, a lot of man-in-the-middle attacks occurred. As noted in the wireless security history section, many groups, schools, and companies produced research papers outlining this vulnerability. The general 802.1x vulnerability is achieved by setting up a rogue access point and getting clients to authenticate to this rogue access point. When these clients connected and passed their credentials, the attacker would use those credentials to connect to the actual network.

When looking at this in relation to current-day wireless systems and clients, one notices that Windows XP has a built-in ability to connect to a network on its own. Service Pack 2 updated the wireless configuration tool on Windows XP and allowed for more manageability for wireless connections. Even with the service pack two, wireless client cards will automatically connect to networks. This means that Windows users are using the default wireless configuration tool and they can easily be swayed off their access point and onto the one an attack controls.

Let us now walk through a real-world example of how to compromise an MS PEAP 802.1x wireless network. An attacker will need to find a target that is using an 802.1x wireless network. For this example, let us assume that the victim is running 802.1x with MS PEAP. This is one of the more common architectures used today because of the cost savings of using an existing MS server, and the savings connected with not deploying a PKI has often led companies down this road.

Getting back to the attack scenario, the first step is to find the network. To do this, an attacker can war drive around until he stumbles upon a network or use some information-gathering technique (discussed previously). For this example, the attacker has his eyes set on a single company. As the first step, the attacker war drives past the building to see if it has any wireless networks. This has proven valuable in letting the attacker know for sure that a wireless network exists at the target location.

The next part of the process involves sniffing the airwaves to find the SSID and making sure that the network is using MS PEAP with 802.1x. An attacker can sniff the network in two ways. One way is to set up a long-range directional antenna such as a yagi. With this, an attacker can sit completely out of sight of the physical security guards. With this, he can point the yagi attached to his laptop toward the building and sniff the airwaves. The other way is much more risky but just as easy. The attacker could just walk right in the door ask the person at the front desk about using the restroom or inquire about a job. All the while, the attacker could have a PDA in his pocket capturing packets and integrating the company's network with a number of wireless tools. Next, the attacker will try to find the SSID. This is easy, although some 802.1x deployments encrypt the management frames. This might pose a problem; however, today, even with certain implementations where management frames are encrypted, all probe requests and probe responses will have a cleartext easy-to-read SSID.

So now the attacker has found the SSID. What is he then going to do? First, he needs to set up a server and an access point to use as a rogue device. The server needs to run on a laptop for portability; it also needs to run the following services: DHCP, DNS, and Web Server. This server could also be set up with one of the many hotspot authentication programs available and will provide this functionality for the attacker in a single software package. Next, the attacker will need to set up the access point to use the same SSID as the target. The access point should have no security on it and be set up just with the SSID and full power. This will allow any wireless device to connect to the attacker's wireless network if the attacker's signal is stronger.

Now that the attacker has set up the access point with the server, he needs to go back to the target company and attach a yagi antenna to the access point. Once this is done, he will point it into the window of the target's building. As more and more people start connecting to the real company's wireless network, some will actually connect to his (the attacker's) access point instead. Once this happens and someone connects to it, a screen pops up asking if they want to connect to this network even though it is not secure. This may turn off the first user and make him think about it; however, given time and the number of users getting this screen, someone will just click on it. It does have a familiar-looking SSID that they are accustomed to. Even if no one connects to it, if someone leaves his PC on the desk, boots it up, and walks away, XP will try to connect anyway. Any of the listed ways to get someone to connect will work. The real purpose is to get a single person to connect and, thus, all that is needed is one person.

Finally, someone connects and associates to the attacker access point. Now that the user is connected, the DHCP server on the attacker's setup will serve up an IP address, default gateway, and DNS. Now the user is going to try to connect to some kind of network service or resource. Maybe he will try e-mail first and nothing will happen; in troubleshooting, he will eventually attempt to access the Internet. The goal is to get someone to access the Web. Although it might not be the user's first attempt, once he connects to the network, he will most likely attempt to go to the Web as part of his troubleshooting efforts before calling the helpdesk. Once the Web attempt is made, a DNS request will be sent to the default gateway for the DNS server. The attacker will take the request and respond with his Web server's IP address as the page the user was requesting. Now the user will be directed toward the attacker's Web server without even knowing it.

Once the Web page opens, a piece of Java script will launch that pops up a window. This window will look just like the default 802.1x authentication windows that the user has seen previously. Now the user will breathe a sigh of relief and type in his username and password. Once he hits the OK button, the form is sent back to the attacker along with the user's full credentials. Now the attacker turns off his access point and the user connects back to his original access point and re-authenticates. Once the user does this, everything works and he (the user) has no idea what happened. To make things worse, he never thinks twice about even placing a helpdesk call because everything is now working.

And now the attacker can break in as a valid user without attempting any password guessing or brute forcing. This will not alert anyone, even with a conventionally wired IDS. The only way to detect this attack would be with a rogue access point detection software suite or wireless IDS. This can be prevented by using EAP-TLS; however, most enterprises are reluctant to deploy PKI across the entire enterprise unless they have some regulation requiring it.

### 13.2.7 Attack on Michael

When Michael came out, a need to prevent message modification attacks was built into it. This method instructed the access point to deploy a feature called the TKIP countermeasure procedure. The details of how this works were identified in the TKIP and MIC section in Chapter 12. As a quick reminder, the process works by having an access point shut off if it receives two MIC failures in less than one second. When the access point shuts off, it will only do so for 60 seconds and then it will come up and require all of its new and previous users to re-key to gain access to the network. An attack was found in which an attacker could send corrupted traffic to the access point. This corrupted traffic would pass the WEP IVC and the frame CRC check; however, when it came to the TKIP examination, it would trigger the countermeasure. If an attacker did this, he could bring down the network until he was located and physically dealt with.

### 13.2.8 Attacks on Wireless Gateways

Many wireless gateways use several means of authentication. This is done to allow the widest versatility in relation to client selection. If the gateway can support a wide variety of clients, it has an edge compared to other non-accommodating wireless security solutions on the market. This led to specific attacks against weaker authentication means. The real threat to gateway systems comes from another strong selling point, being clientless. To have a gateway clientless with the ability to support a wide variety of operating systems, SSL is often used. When SSL is used, the gateway prompts users with a default Web site that they must authenticate to before they can gain access past the gateway. Now one look at how attackers have undermined their security.

Some tools are available to allow for an SSL proxy and can defeat the security of wireless gateways. The first step in performing this attack is to set up a proxy server and place it between the gateway and the client. When the client fires up its wireless connection, it will connect to the proxy and then the client and the proxy will establish an SSL connection. After this takes place, the proxy will establish another SSL connection with the wireless gateway. Once both SSL connections are created, the wireless user or the victim will see the wireless gateway and authenticate to it. This authentication, unbeknown to the victim, is susceptible to eavesdropping. The normally encrypted authentication traffic is briefly decrypted at the proxy server and then is re-encrypted and sent to the gateway. One piece of software that can perform this attack is called Achilles; it is available from DigiZen security group at http://www.digizen-security.com.

Another interesting attack on clientless gateways is the same as in the 802.1x section. Creating a rogue access point and network to look like the gateway can lure clients into authenticating to it. This would be set up the same way, with the only difference being a Web site that looks like the gateway device. Once the first user connects to this attacker-owned Web site, his credentials can be stolen and used to access the network.

### 13.2.9 Attacks on WPA and 802.11i

On November 4, 2003, Robert Moskowitz found out that WPA and 802.11i suffer from an offline dictionary attack. This is only true when preshared keys are in use. This is the result of an information exchange required to create session keys in WPA and 802.11i called the four-way handshake. The 802.11i section discussed how session keys are created from master keys. This process involves taking the master key, two nonces, and both sender's and receiver's MAC addresses as input into an algorithm. When preshared keys are used, the creation of this master key is achieved by passphrase, SSID, and SSID length. These fields are put into a PBKDF2 algorithm, which performs a hashing function 4096 times, creating a 256-

bit key. For an attacker to perform this same operation, he would have to know the SSID, SSID length, and passphrase. Looking at all of these pieces of information, one can see that the only truly confidential item used in the creation of a master key is the passphrase. The SSID as well as its length can be easily captured by wireless sniffing software. With this information, an offline dictionary attack is possible.

So to perform this attack, the SSID must be identified with the use of a wireless sniffer or through some other means. Once accomplished, an attacker must observe the four-way handshake used to create the session keys. In the second message of this process, an EAP message is sent that contains two values PTK and KEK that are hashed using MD5. This hash allows an attacker to try multiple passphrase combinations to find one that matches.

# 13.3  Access Point Compromising Techniques

Thus far, the discussion has focused on attacking the security mechanisms of a wireless network in an effort to provide access to the network. There has been no mention of attacking the access point itself. Most networking equipment has a number of vulnerabilities that exist. Some are just related to the use of certain protocols and some vulnerabilities lie within particular vendor devices. This section looks at several attacks and threats that face the access point itself.

## 13.3.1  Remote Management Attacks

As with most network equipment, there are many ways to connect and manage access points. Some of these methods are considered non-secure and must be carefully evaluated for each situation in which they are used. To understand what threats relate to each type of management, we outline the most common methods to connect, change, monitor, and control access points. Using this information, one can make educated decisions on what types of management should be allowed and what types should be restricted.

### 13.3.1.1  Telnet

The first remote management protocol to look at is telnet. Telnet is used to perform remote terminal sessions on devices. It allows an administrator to configure networking devices from anywhere there is IP connectivity. Chapter 10 outlined how to connect to some access points via telnet. Telnet is an old protocol created in the 1960s. Because it was created so

| 6 | TELNET | .. | |
|---|--------|----|----|
| 4 | TELNET | .. | |
| 2 | TELNET | Password: | |
| 6 | TELNET | .A....,S= 138010964,L= | 0,A=3471540585,W=17424 |
| 8 | TELNET | c | |
| 1 | TELNET | .A....,S=3471540585,L= | 0,A= 138010965,W= 4075 |
| 2 | TELNET | i | |
| 0 | TELNET | .A....,S=3471540585,L= | 0,A= 138010966,W= 4074 |
| 6 | TELNET | sc | |
| 1 | TELNET | .A....,S=3471540585,L= | 0,A= 138010968,W= 4072 |
| 4 | TELNET | o | |
| 3 | TELNET | .A....,S=3471540585,L= | 0,A= 138010969,W= 4071 |
| 0 | TELNET | | |

**Figure 13.1   Telnet sniffer trace.**

long ago, many security features were omitted. One of the major features is the encryption of authentication traffic. This means that all authentication traffic performed inside a telnet session is done so in cleartext. That is, if an attacker wants to eavesdrop on a telnet session with a network sniffer, they can see the administrator's username and password. In Figure 13.1, one can see the output of a network sniffer recording a telnet session with an access point.

### 13.3.1.2 HTTP

Another rather easy way to administer an access point is through a Web browser. Some access points only allow this type of administration. This is commonly the easiest way of setting up an access point. HTTP has always had some security concerns. Most HTTP pages are vulnerable to an automated password guessing attack. This can be done by attacking with words known in a dictionary or trying every combination in a brute-force attempt. Some older Cisco VxWorks access points suffered from this attack, although the new IOS code has created a mechanism that stops most of the common HTTP attack tools from operating correctly. Almost all Linksys access points are vulnerable to this type of attack. Because of this attack, large amounts of Web-based interfaces have been disabled on networking equipment in many companies. Using this type of administration is more user friendly and easier to work with, although with that ease of use comes an increased risk of attack. Weighing usability against security is a task that faces all security personnel.

### 13.3.1.3 RADIUS

Many times, a company will realize that it has too much networking equipment and can no longer keep the simple username and password list on the device. This is when an AAA solution should be deployed.

This AAA solution often uses RADIUS; if that is the case, most of the exploits and vulnerabilities listed in the RADIUS section above apply not only to wireless authentication, but also to management access to the access point itself. Because of flaws in how RADIUS sends its messages, many tools have come out that can capture a RADIUS message and extract a key. This extraction is limited to time and resource power, although it can be done. If someone were to log in to an access point and the access point was set up to use RADIUS, the conversation between the access point and the RADIUS server could be captured and later deciphered into the user's name and password.

### 13.3.1.4 SNMP

Simple Network Management Protocol (SNMP) has been around since 1988 when RFC 1067 was created. Since then, SNMP has gone through RFC 1098 in 1989 to where it is today in RFC 1157. The current RFC 1157 was created in 1990. This is similar to telnet in the fact that the RFC was created before security was a major factor in any standard development. All three RFCs are SNMP Version 1. SNMP has been through three versions. Versions 1 and 2 lacked much-needed security features. SNMP Version 3 was approved by the Internet Engineering Steering Group (IESG) as a full Internet Standard in March 2002. Most all equipment supports Version 1; some equipment has started to utilize the other two versions. The discussion here focuses on Versions 1 and 2 of SNMP because they are the most widely used.

SNMP is used to remotely monitor and apply changes to networking devices. It can be a valuable tool to troubleshoot and trend network usage. It is a very powerful tool. This power can be good — if it is in the hands of an administrator — or bad — if it is in the hands of an attacker. With SNMP access, an attacker can perform almost limitless configuration changes. SNMP has some very basic security mechanisms. This security is based on two strings: one that allows for read access and one that allows for write access. To make any changes, one must have the write string. These strings are similar to passwords; if they are known, then the corresponding actions of read or write can take place.

If an attacker had access to the write string, they could reboot a device, change the configuration of a device, or remove an existing password. For an attacker to get these strings, some actions must be performed. Somehow the attacker must get a hold of these strings. This is where a number of SNMP attack tools come in; with one of the types of available SNMP tools, an attacker can try a dictionary or brute-force attack on a device. This will reveal the SNMP string and give the attacker access to the device. Once the write string is found, the attacker can reset the password via SNMP and have full access to the device. If only a read string is present, a look at the configuration could yield the password. Sometimes these passwords are encrypted; if this is the case, there are many tools available to decrypt them.

## 13.4 Chapter 13 Review Questions

1. What is the process of scanning devices for open ports called?
   a. Information gathering
   b. Enumeration
   c. Compromise
   d. Expanding privileges
   d. Cleaning logs

2. Using a corrupt EAPOL packet creates what type of attack?
   a. Cracking
   b. Man-in-the-middle attack
   c. Denial-of-service attack
   d. Information gathering

3. Why would an attacker use another wireless technology outside what the target is using to place a rogue access point?
   a. To get a better connection
   b. To provide better bandwidth
   c. To conceal the device easier
   d. To prevent rogue detection software from finding it

4. What was the most significant attack on 802.1x?
   a. Cracking
   b. Man-in-the-middle attack
   c. Denial-of-service attack
   d. Information gathering

5. When an attacker sends a piece of information that is encrypted back into the network in hopes that the access point will decrypt it, what attack is performed?
   a. Dictionary attack
   b. Plaintext cryptanalysis attack
   c. Double encryption attack
   d. Shared key authentication attack

6. SNMP can be used to reset passwords.
   a. True
   b. False

7. When using RADIUS, how many characters should the shared secret be?
   a. 32
   b. 8
   c. 19
   d. 24

8. 802.11i with pre-shared keys provided no current security risks.
   a. True
   b. False

9. What attack on WEP are most WEP cracking tools based on?
   a. Dictionary attack
   b. Plaintext cryptanalysis attack
   c. Double encryption attack
   d. Shared key authentication attack
   e. Stream cipher attack

10. Telnet provides a secure means of remote authentication.
    a. True
    b. False

11. When using a wireless gateway, what protocol is most likely to become compromised?
    a. IPSec
    b. AES
    c. SSL
    d. EAP-TLS

12. The process of using Google to locate relevant data about a target is performed at what step of the hacking process?
    a. Information gathering
    b. Enumeration
    c. Compromise
    d. Expanding privileges
    e. Cleaning logs

13. Name two ways an SSID can be captured.
    a. Capturing probe frames
    b. Capturing beacon frames
    c. Capturing authentication frames
    d. Capturing association frames

14. To compromise a key, which of the following are needed? (Select two)
    a. Cleartext packet
    b. WEP management frame
    c. WEP encrypted packet
    d. Wireless client

15. What would be the most likely place to find out about a company's IT equipment?
    a. Company Web site
    b. War driving
    c. Checking help forums for a company's name
    d. Using Google

## Chapter 14

# Wireless Security Policy

This chapter explores the process of creating a wireless security policy. In this process, we address each of the steps needed. The first step in creating a wireless security policy is to perform a risk assessment. This assessment will inform the corporation about all the risks currently existing in their organization. After properly examining the results of this assessment, they are used as a starting point to understand what needs to be addressed and in what order of importance. This will lead into the policy-writing process. After the policy-writing process is finished, this chapter discusses the different sections of a wireless security policy and added details, ideas, and lessons learned from each of them that apply directly to wireless.

## 14.1 Policy Overview

When developing a wireless security or any policy, a well-defined process must be understood and followed to ensure that the policy created is strong, correct, and has the backing of senior management. When all of the parts of this process come together to form the policy, it must address the corporate guidelines around the use of wireless technologies and have the enforcement needed to uphold its contents. This means that senior management must be willing to enforce the rules of this document and champion it throughout the process. Without this senior-level involvement, the policy could fall onto deaf ears.

Before proceeding into writing a policy, some terms and definitions should be fully understood. All too often, companies create documents that traverse the lines between policies, procedures, standards, and guidelines. Let us take a quick look at each of these terms and what kind of document scope is involved with each of them. This will help us understand where certain requirements go and why some documents should be exact when others should be vague.

When discussing each of these documents, one should know why an organization needs to create multiple documents to address a single item such as wireless access. One point, which is often a reason why policies are created, is compliance. Some companies have requirements to comply with certain laws or regulations sanctioned by various government agencies. Another reason behind this is confidentiality. Often, policies have strategic information inside them that should not make its way to contractors who may only need a standard document outlining how to perform a certain task. One of the major reasons behind dividing up these documents is the life-cycle difference of each document. The bigger a company is, the more difficult it becomes to get the correct officers or senior leaders together to approve a document or a version of an existing document. On the other end of the spectrum, certain changes are frequently made to a standard document outlining changes in technology, hardware, software, vendors, or lessons learned. For these changes, amassing officers and senior members of an organization is overkill. Because of these reasons and many others not outlined in this book, creating a tiered documentation approach is essential to operating any company in the most effective manner.

### 14.1.1 Policies

Policies are high-level documents created to represent the strategic thinking of senior management. They are clear, concise views of where senior management wants to take the company. They do not describe the details of how the company is to get there but, rather, they lay the groundwork by telling the organization what the goals and objectives are. It is left to another document to outline the exact details of how this will be accomplished.

Let us use the analogy of a roadwork project to describe how each of the terms interacts with each other. The policy document would be the driving force, saying that we need a road in this location. The document would state the basic goal from above. Inside the document would be comments or requirements about the road. This would be similar to saying that the road must have the ability to support the current population as well as any anticipated growth within the next ten years. It could also sate that the road must be finished by a certain date or started on a certain date. This policy document does not dictate how wide the road will be or if it will be concrete or asphalt. It only points out the goals and objectives as well as the timing expected to achieve these goals or objectives.

## 14.1.2 Standards

Standards set the *how* in relation to a policy document. After senior management has developed a policy outlining its intended goals, it is left to a standard document to outline how the organization is going to achieve them. The standards document outlines the details such as the selection of hardware, software, or technology. This document would go into detail about how each piece of hardware should be configured. It would also address versions of software or firmware required on the hardware. It would outline how a specific technology will be leveraged.

Returning to the roadwork analogy, one sees that a policy was put into place stating that a road must be created. The standards document will outline if the road is three or four lanes wide, if it uses concrete or asphalt, and how many signs, lights, or exits it should have. Got the idea? To recap here, the standards document defines the *how* in relation to the policy document, which outlines the *what*.

## 14.1.3 Guidelines

Guidelines are similar to standards in the sense that they also detail the *how*. Guidelines take a different approach by not saying hard and fast thou must, but rather that thou should. This is a common approach in organizations where departmental management does not agree. In this situation, a guideline can be created to meet the goals of a policy, although there is no requirement to follow a guideline. Another use of guidelines is to leave certain options up to each division or group while still making certain requirements in a standards document. In large organizations, senior leaders often use a management technique in which they let each group try to implement a corporate initiative in their own way to see which one did it the best. In this approach, a standards document is a way to keep certain requirements static across the organization while allowing guidelines to be created for each group.

Going back to the roadwork analogy, a guideline would be used to say that the way Group A built their section of the road is this. If Group B wants to use the same method it can, or it can create its own. These guidelines are set up so that if Group A's management and Group B's management do not get along, a power struggle does not impede the project. Because guidelines recommend and do not require, they allow different groups that have different ideas the ability to follow or discard them. Where a standard plays into this is that a standard could stipulate that all groups must use concrete and must create four lanes. The guidelines as to how to do this would not be required, unlike in this case.

### 14.1.4 Procedures

Procedures outline specific tasks that are defined in a standards or guideline document. They are at the bottom of the document list as far as authority and scope. Procedures apply to the small sections of standards. They are primarily employed as working installation guides used by the resources physically doing the work. A procedure only defines a task detail about a certain section of the overall standard or guideline.

Looking at the roadwork analogy, the procedure would be a document stating how to assemble a streetlight or how to mix the concrete. It would be used by the lowest management level of an organization, detailing to them how their subordinates should perform their job functions.

## 14.2 The Policy-Writing Process

Writing a security policy involves many steps. The first step required is a risk assessment. A risk assessment is key to creating a sound security policy. Without knowing what level of risk is facing an organization, creating a security policy to combat this risk becomes almost impossible. After a risk assessment is performed, some general guidelines relating to what risks are most important must be set by upper management. Upper management's involvement in a security policy is required or its substance will never be achieved. If people at the top are not behind a policy, all their subordinates will also be unwilling to accept the policy.

After upper management has listed each risk and assigned a priority, a team needs to work up a draft policy. This policy will be created by examining each risk in order of importance and creating controls to reduce each of those risks. Sometimes, the risk may warrant transferring it to another party; this is often done by the purchase of some form of insurance. After creating all the controls, the next required step involves socializing and approving. In this step, other groups, departments, or business units will read and give their comments on the draft. This is required because something in the draft might be unachievable or unrealistic. An example of this would be to require the use of SSH instead of telnet for all network management. Once the network architect sees this, he may raise a big red flag because some of the equipment does not support this. This could mean that if the policy was put into place, that a large investment would need to be made to the network to accommodate the policy. Things like this must be worked out before the next step (approval) takes place.

Once there are a number of groups behind a policy, they can get approval from upper management. Upper management is more likely to say yes and approve something that has been accepted throughout the company. This could go either way; just having a policy generally accepted by multiple groups across an organization will help it become approved, although it will not ultimately affect its position. Once upper management approves the policy, a final copy is created and handed over to an operations team so that it can be put into place.

## 14.3 Risk Assessment

The first step in initiating any security policy is a risk assessment. A full, correctly performed risk assessment will identify what risks are present today as well as what current controls are working correctly to reduce that risk. This assessment could take a wide scope, examining all risks facing an organization, or a specific scope such as all risks related to wireless technologies. From this information, controls or actions made through a policy can be implemented to offset the risk previously identified. An assessment will give a policy-writing team an idea of what is considered a major risk and what is considered an acceptable risk. This will ensure that the policy-writing team dose not waste time developing or improving on already good controls.

The terms and strategies that define a risk assessment vary, depending on who is performing the assessment. There are a number of approaches used by companies to perform a risk assessment. Insurance agencies have played a major role in risk assessment. Their livelihood rests on the fact that they can properly gauge risk and provide mitigation to risk for their clients. When a company buys insurance, it is transferring the risk that it perceives as high to someone else. This means that if the threat were to occur, the company would not have to absorb the risk alone; some of the burden would be shifted to the insurance company.

When a risk assessment is performed, the key goals are to:

- Identify the threat.
- Understand the impact if the threat occurred.
- Understand how often the threat might occur.

Once one has a full understanding of the threats, how much they cost, and how often they occur, one can then start to make an assessment from this data. There is a time-tested approach to performing this evaluation of risk. It uses common values and formulae to create a realistic portrayal of risk in any given enterprise. In the industry, this formula is the result of the following terms: EF, ARO, SLE, and ALE. Each of these terms is detailed below.

### 14.3.1 Exposure Factor (EF)

This is the impact on the value of an asset. It is often expressed as a percentage and can range from 0 to 100 percent. This percentage is the amount of loss that affects an asset. If an asset is a trade secret and a competitor gets hold of it, the EF value would be 100 percent. This value is necessary to create what is called single loss expectancy (SLE). This helps to complete the formula that gives the annualized loss expectancy (ALE). Both terms (SLE and ALE) are defined below. For now, understand that the exposure factor EF is the percentage of loss that occurs to an asset if a risk occurs.

### 14.3.2 Annualized Rate of Occurrence (ARO)

This term describes the frequency of threat to occur over a single year. This value is also used to create another term, called the annualized loss expectancy (ALE). The annualized rate of occurrence (ARO) can be represented properly even if an event is only likely to occur every ten years. That is, the ARO is 0.1, or the likelihood that this event will occur is one in ten.

### 14.3.4 Single Loss Expectancy (SLE)

Single loss expectancy (SLE) is a dollar value used to measure the impact a threat might have if it occurred. The formula used to gauge this value is an asset's value multiplied by its exposure factor. This will equal the single loss expectancy (SLE). This value is used to show how much a single impact of an identified threat would cost in dollars. Single loss expectancy (SLE) is calculated as follows:

Asset Value × Annualized Rate of Occurrence = Single Loss Expectancy
$ \qquad\qquad ARO \qquad\qquad\qquad SLE

### 14.3.5 Annualized Loss Expectancy (ALE)

The annualized loss expectancy (ALE) is a value that determines the amount of money a threat might cause in a given year. This value is made up of the single loss expectancy multiplied by the annualized rate of occurrence (ARO). This value can help classify a number of identified risks using the ALE of each one compared to the other. Most events have widely varying rates of occurrence. When looking at the entire cost a risk might bring to an organization over the course of a year, it is often noticed that a small risk that frequently occurs can cost more in a year than one large risk that carries a high cost. This ALE is calculated as:

Single Loss Expectancy × Annualized Rate of Occurrence =
SLE ARO

Annualized Loss Expectancy
ALE

An example of this involves comparing two risks. Risk 1 deals with incorrectly filing a form. This risk occurs, on average, once a month and costs $1000 each time it occurs. Risk 2 is the loss of a server resulting from a hardware malfunction. This might happen once in two years and cost approximately $10,000 each time it occurs. When one calculates the ALE of both risks, one sees that the smaller risk of incorrectly filing a form actually costs more than that of a server failure. This is why it is important to perform the ALE operation to identify multiple risks correctly in relation to each other.

Given an understanding of the terms used to create a risk assessment, one now needs to understand the financial outcome of a threat. A lot of companies and senior leaders often say that this approach to risk management is as good as guessing. This is true in most aspects because of the guesswork that goes into creating the amount of financial loss that a threat might impose. Because of this, many people and companies tend to lean away from this hard-dollar quantitative risk assessment approach and more toward a process that does not try to factor a dollar amount into specific risks. When performing a risk assessment using the financial information to illustrate risk, one has performed a *quantitative* risk assessment.

The *qualitative* risk assessment process, on the other hand, uses a rating scale to list risks in order of importance based on the amount of damage the threat might impose. This approach still includes most of the terms used in the quantitative approach, although dollar amounts are never part of the formula. Some quantitative approaches might use dollar amounts gathered from performing ALE formulae to gauge importance; however, these dollar amounts will never be shown in the assessment. They will only be used for prioritizing risk.

# 14.4 Impact Analysis

The impact analysis process is used to understand what impact a change might induce. When taking on risk to update a system or change a hardware platform, it is sometimes cost effective to have a full understanding of the amount of risk and potential loss that the change can cause. This step is primarily a risk assessment although its scope is to deal with a specific event or change outlining how it could negatively affect a business.

An example of an impact analysis would be the upgrading of a critical application. This analysis would look at the amount of loss that adversely affects the organization if the upgrade did not work and the system shut down. This analysis would provide management with the information needed to make decisions on how to perform the upgrade. This impact analysis would help one understand that any added research should be justified to offset the risk of failing.

# 14.5 Wireless Security Policy Areas

Given a general understanding of security policies, one now needs to get more in line with the details of a wireless security policy. This policy has its own areas that need to be outlined and addressed outside of other information system policies. It should encompass all wireless technologies inside an organization. With the lines rapidly blurring between phones, laptops, and PDA devices, a policy covering all wireless communications will help prolong its life cycle and adapt more easily to newer technologies. A wireless security policy should be present in every company, including those not using wireless networking technologies. Having a wireless security policy in a company not using wireless networking technologies enforces the fact that certain wireless technologies are not allowed. This sets the expectation of what will happen to employees if wireless communication technologies or methods not approved in the policy are being used.

Policies are difficult to write for companies that already have in-depth policies. In this environment, many of the basic policy points are addressed in their own documents. For example, most companies have a password policy. In a wireless security policy, passwords are often used. To keep everything from being contradictory, two actions can be undertaken. First, one can update all general policies that wireless can affect; this means updating the password policy, acceptable use and abuse policies, and many others. This makes the wireless security policy scattered across many other policies, thus making it difficult to update and maintain.

The second option is to create subsections in the wireless security policy. These subsections will address each of the areas where there are other policies already in place. These other policies can be referenced or left out completely. As long as no contradictions take place, keeping these subsections inside the wireless policy will help in the maintenance of this policy. Each company will need to make its own decisions about where and how to address each area of wireless security policy.

Once the risk assessment is complete and upper management has relayed its top priorities for the policy, the writing portion begins. The next few subsections address the main areas encompassed inside a wireless security policy. Before getting into each area, a level set must take place in the policy stating what it is, what it is for, and whom it affects. In some large organizations, policies are often not companywide. When this is true, there must be an explanation of the area(s) in which they are and are not in force.

## 14.5.1 Password Policy

The usage of passwords has been part of information security for its entire existence. In this existence, there have been many uses for passwords and many ideas about how to create strong passwords. This subsection discusses how passwords are used and what has been done to create strong passwords. This information will be used to help adopt a happy medium between strong passwords and the feasibility of clients remembering these passwords. This medium will ensure strong passwords that are not so complex that they must be written down.

Before getting into the password policy section, let us briefly mention the threats related to the passwords themselves. Password cracking tools can attempt multiple password types until one is granted access. To understand how these work, it is necessary to cite each kind and detail their functions. All password-attacking tools perform a guess on a password. This guessing can be words that are part of a word list or dictionary. This is called a dictionary attack. If a password does not have a dictionary word, a brute-force attack can be performed as well. A brute-force attack will exhaust all possible combinations of characters. This test usually is time and resource restrictive, meaning that without the right equipment and time span, it may never find a correct password.

The first item that comes up when generically talking about passwords is the concept of factor identity. A password is considered a single-factor authentication mechanism. The factor concept consists of three main categories: (1) something you know, such as a password; (2) something you have, such as token or private key; and (3) something you are, such as fingerprint or retina scan. Using more than one factor makes your risk reduce dramatically. When deciding about adding additional factors to your password policy, remember to take into account that passwords are the least expensive means of authentication. Using another factor on top of passwords or in the place of passwords will increase security, although it will also add expense. Tokens or readers, along with the support costs of the added systems, may prove more expensive than the added security provided by these systems.

When looking at what can be done with password policies, one needs to address password complexity. This complexity is an equalization technique used to find the strongest passwords without causing excessive password recovery issues. Using complexity requirements will ensure that passwords are created strong from day one.

A complexity requirement involves the use of letters, numbers, or special characters. When developing this requirement, a decision must be made as to how many of the three kinds of characters will be required. Requiring two or more of these complexity requirements will ensure that there are no passwords created from words found in a dictionary. Requiring all three will create a password that will require a long, exhaustive brute-force attack to reveal its context. The more complexity one adds produces a negative effect. This effect is seen in an increase in password reset calls to the help desk and, even worse, the sticky note password on a workstation monitor. Requiring very complex passwords usually leads to users writing them down or forgetting them. A weak password is better than a strong one if the complexity makes users write it down and place it in front of their workstations.

Another interesting method used to create and remember passwords is the keyboard layout method. This is often taught to people with a high level of access to secret documents. It works rather well for governments due to the fact that the user actually does not know the password; rather, the user is familiar with a certain set key layout. This gives a person the ability to take a polygraph and pass by honestly saying, "I do not know the password." This is because the user actually does not know the password; rather, he knows how to type it. To get into more detail, this method uses keys that are grouped together. This would be as easy as typing 4rfv5tgb. As one can see, it has complexity and is somewhat random; yet if one were to type this on a keyboard, one would instantly notice how easy it is to remember the keystrokes and not the password itself. One note about this type of password is that most password attacking tools perform a check of closely grouped keys to find passwords similar to this. This may sound a little scary; however, this step is often just before a brute-force attack. With a brute-force attack and time, any password can be recovered.

## 14.5.2  Access Policy

This is one of the most important pieces of a wireless security policy: who can access the wireless network and with what devices. This sub-section addresses the immensely important issue of access. To start, one must understand how the risk assessment addresses different levels of access. What risk rating was given to non-standard, non-company-issued devices connecting to the wireless LAN? What risk rating, if any, was given to standard company-approved and deployed wireless devices? These ratings are often in direct relation to the current wireless network security architecture. For example, if the network is running plain WEP, then both non-approved and approved devices pose major a risk. If the network has a security solution in place beyond WEP, such as 802.1x, this may make non-approved devices unable to connect to the network, thus increasing security and lowering risk.

Now that one has knowledge of the risks facing access, one can look at another form of access: rogue access points. These can be placed by hacker or employees. In either case, each one needs a clear explanation of the consequences and actions team members must take if a rogue access point is identified. This could be as simple as locating and removing any rogue access points. It could also be as dramatic as firing the person who installed it. We have just gone from the least extreme to the most; when writing the policy, management must give input as to how forceful this portion of the policy should be. After the proper risk assessment is performed and management states that removing rogue access points is one of its top priorities, then one can bet that the access policy section will need as many "teeth" as possible. If this is the case, it could easily be as extreme as firing the person who did it. No one wants to fire anyone, so in most cases this is used as a tool to spread awareness. It may have to go as far as making an example and actually firing someone to speed up the process of spreading awareness about the access policy.

This section of the policy can also address what can and cannot access the network. Some devices, such as personal PDA units with built-in Wi-Fi may be a real threat. There could be certain portions of a wireless network running mission-critical application data. This is common in manufacturing where parts tracking and delivery to line is mission critical. It can extend to automatic guided vehicles, which are robots moving parts, units, or even cars through the assembly line. In these special cases, making sure there are no other wireless devices transmitting on the network is critical. Just imagine someone walking onto a plant floor, downloading a large file, and once they roam on the same access point feeding time-sensitive, mission-critical devices, it stops. That can really affect production and waste large amounts of money. This can also be seen in hospitals where patient vitals are often carried over a wireless network. No one would want the ability to download music if it meant their heart monitor would not work correctly.

This policy should directly state what devices are allowed. This can be a list or a business rule stating that a certain management level must approve the device before it can connect to the network. Also, this section should state what could happen to anyone who circumvents this policy and connects an unauthorized device or sets up an unauthorized network.

## 14.5.3 Public Access

Public access is something new to security policies. Some policies have set public access policies in the past, but only a few. For example, a public library may have offered network access before wireless became widely used. With the small number of companies performing public access, policy development is often a forgotten piece.

Today, hotspots are everywhere and most companies that directly deal with people are starting to look at offering public Internet access as an incentive to get people into their locations. Detailing this in the policy will ensure that the public access offered does not negatively influence the company. The policy should define what steps are needed to be able to provide this and what can be done to make sure the company's network is secure. There is also the issue of using hotspots to connect to a company network. Most hotspots have little or no security. Making sure a hacker does not steal or piggyback off a remote user's laptop is very important in protecting the organization and lowering risk. Also, protecting the company's resources on wireless devices is a major issue that must be addressed in a public access section.

Now let us address companies that would like to set up hotspots for customers. These networks should be separate from the corporate network and employ some security features to prevent a connection from jumping into the company network. This is where a policy stating certain high-level goals should be communicated. For example, the policy could state that the hotspot connection needs to be physically disconnected from the company network and use a separate ISP. It could also require the use of a certain security architecture that lowers the risk and does not have the extra cost of a separate Internet connection. Another way to look at it is to disallow wireless access for employees. This may make some employees angry and lead to rogue access points. One way to fix that is to design the hotspot network for both customer use *and* company use. Company users could use a VPN to connect to the company resources.

This would be the same as if they were using another hotspot down the road. If this approach is used, the policy only needs to address employee wireless access, regardless of where the access is originating from. This is because access could be through a VPN or through the same VPN at a hotspot. Whatever route a policy may go, it is important to understand that there is significant risk involved in running a hotspot.

The next portion of this policy needs to address the risks cited in the wireless end device section. When a proper risk assessment is performed, a number of risks may come out that do not directly relate to wireless networks. Some technologies, such as e-mail retrieval and remote management, have common flaws. These flaws are considered acceptable in the situation where someone is connected to a wire that is connected to the Internet. The Internet can be a wild place where traffic can float around the globe, although the only likely candidates that would be able to sniff one's network traffic are employees of service providers. That risk is often considered actable when these insecure technologies were evaluated and deployed. Now that there is wireless where anyone can see the traffic, different rules and reevaluations should be undertaken on these insecure protocols.

Now to deal with someone who would steal company information from a wireless device. The next subsection (Section 14.5.4) of physical security addresses the theft of the device itself. This section, however, addresses the data separately because there are many ways of stealing the data from a device without having to steal the device itself. If one looks at using wireless technologies such as Bluetooth or Wi-Fi for hacking, one can see from Chapter 13 that many ways exist to compromise wireless devices. Protecting the data on the device is extremely important. Many PDA devices carry contacts and sales information that is critical to companies. If this data falls into the wrong hands, a business could quickly be out of business. To protect this data, some steps must be set forth in the policy. One might be to use password protection on all wireless end devices. Another would be to disallow creation types of communication technologies not considered secure.

## 14.5.4 Physical Security

The physical security section has its own unique topics. One of them deals with the wireless network itself. How many times have you seen a wireless network? Next time you are in any retail store, look around and you will see them. You will see the antennas hanging; sometimes you will see the access points themselves. In this case, the access point and antennas are located well out of reach. In areas such as office buildings and hospitals, the access points are often located in the hallways where everyone can see them and touch them. This is a physical security risk and controls must be incorporated in this policy section. Another common physical security issue is that of the end devices themselves. These devices are small and often forgotten or stolen. Making sure that they are safe from thieves, and the data on them is safe as well, is an important part of this policy. This chapter section looks at each of the risks that directly relates to wireless physical security and how certain policy steps can mitigate or prevent them altogether.

The first risk comes from stealing the access points themselves. Some common commercial access points range from $100 to more than $100,000. Keeping them safe is important. Today, almost everyone is starting to experiment with wireless networks at home. This means that the threat of someone walking off with a wireless access point is very likely. This risk is even greater if the facility is open to the public, such as a hospital. To protect against this, certain rules should be set and these rules can be set in this section of the policy. Making statements like, using enclosures to house access points can reduce this risk. Another approach would be to require that the access points be mounted out of sight. This would be as simple as installing them above a ceiling tile. Their antennas would still be located below the tile; however, they are less likely to draw as much attention as an access point with highly visible flashing lights.

The next portion of the policy relates to wireless end devices. As stated above, these devices are easily stolen or lost. To help prevent this, some steps must be taken. For example, do not leave company assets unattended. One step that can be taken to mitigate this risk would be to use software that can send cellular signals to these devices. These signals can erase the device and lock it from use. This technology is very helpful if a wireless device is stolen or lost. Education is the key to preventing this from happening. If proper steps and technology are undertaken, correctly safeguarding these devices will lower the risk facing them.

## 14.6 Chapter 14 Review Questions

1. What is the first step in developing a wireless security policy?
   a. Write the policy
   b. Perform a risk assessment
   c. Security audit
   d. Impact analysis

2. What process results in the identification of flaws in the wireless network?
   a. Write the policy
   b. Perform a risk assessment
   c. Security audit
   d. Impact analysis

3. What types of companies should have a wireless security policy?
   a. Companies with trade secrets
   b. Companies that have wireless networks
   c. Companies that do not have wireless networks
   d. All companies

4. What is the most important factor in creating a wireless security policy?
   a. User buy-in
   b. Management buy-in
   c. Grammatically correct content
   d. Correctly performed audit

5. Which document does not go into detail on how to implement or deploy a solution?
   a. Standards
   b. Procedure
   c. Process
   d. Policy

6. Which risk assessment model has a cost involved?
   a. Quantitative
   b. Qualitative
   c. ALE
   d. SLE

7. A password alone is considered _____.
   a. Weak
   b. Single factor
   c. Two factor
   d. Smart card

8. What is needed inside a wireless security policy to protect users from the threats of hotspots?
   a. Strong ciphers
   b. Strong usage policy
   c. Public access policy
   d. Acceptable use policy

9. Trying all types of password combinations is called a
_____.
   a. Dictionary attack
   b. Brute force attack
   c. Random attack
   d. Combo attack

10. Which password has three complexity requirements?
    a. hfdj942
    b. 23j@w09
    c. 1q1q1q1
    d. Password

11. Which action will not help the strength of a password?
    a. Limit the frequency of reuse
    b. Increase the complexity
    c. Increase the length
    d. Change weekly

12. What document is used to instruct the lowest level of management?
    a. Standard
    b. Procedure
    c. Process
    d. Policy

13. What document is used to define *how* a policy is achieved?
    a. Standard
    b. Procedure
    c. Process
    d. Policy

## Chapter 15

# Wireless Security Architectures

In progressing through this book, one has seen the different security methods used over time to secure wireless communications. One has also seen how some of them have fallen victim to oversight or incorrect security implementations. This chapter puts those security methods together to see a number of secure wireless solutions. These solutions are used to combat the risks involved in wireless communications. Before getting into each of the detailed architectures, there are some security basics not outlined in early chapters that should be discussed. One needs to understand the different encryption methods and their corresponding ciphers. This information will help to set the baseline for comparison.

When discussing securing wireless, one needs to understand how to create different wireless security architectures. This chapter sheds some light on the many different approaches that have been used over time to address the risk involved in deploying a wireless network. This chapter discusses each of them, exploring the risk, pros, and cons of each architecture. This information will help provide a better understanding of how to deploy a secure wireless network in a number of different scenarios.

The four high-level architectures are:

1. Static WEP
2. VPN
3. Wireless firewall or gateway device
4. 802.1X

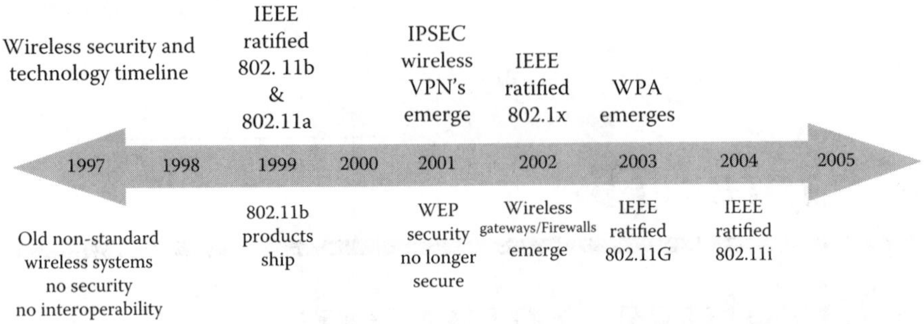

| Wireless security and technology timeline | | IEEE ratified 802. 11b & 802.11a | | IPSEC wireless VPN's emerge | IEEE ratified 802.1x | WPA emerges | | |
|---|---|---|---|---|---|---|---|---|
| 1997 | 1998 | 1999 | 2000 | 2001 | 2002 | 2003 | 2004 | 2005 |
| Old non-standard wireless systems no security no interoperability | | 802.11b products ship | | WEP security no longer secure | Wireless gateways/Firewalls emerge | IEEE ratified 802.11G | IEEE ratified 802.11i | |

**Figure 15.1  Wireless security architecture timeline.**

In the different scenarios, we are going to look at different size environments (from 10 users to 7000 users): (1) a small network consisting of one site with one to ten users, (2) a medium network consisting of two sites with five hundred users, and (3) a large network consisting of ten locations with a total seven thousand users. Using these different levels of size will help adapt any of these architectures to a large number of companies

The aim here is to be as generic as possible. Many vendor products can be used for analysis, although only a select few are examined in this chapter. In the past, we had different limited approaches to take based on when in time we were deploying a wireless network. When a need for wireless security was realized, certain options were not available at certain times. Using the timeline in Figure 15.1, one can see that if one had to build a wireless security solution in early 2001, there would not have been any 802.1x solutions available. In addition, if one were to build a wireless network today, one would be able to leverage 802.11i because it is now published. By examining these various architectures, one can gain valuable knowledge about how most wireless networks were designed, as well as how some of the newer wireless networks are designed.

# 15.1  Static WEP Wireless Architecture

The first architecture is static WEP. This is what the majority of wireless deployments are running today. Most companies are trying to change this into one of the other, more secure architectures. Still, WEP has some merits, such as speed and its standardization. A lot of old wireless equipment is only able to support WEP and WEP only. WEP is located inside a number of 802.11 standards; this has led to wide adoption of

Wireless WEP
security architecture

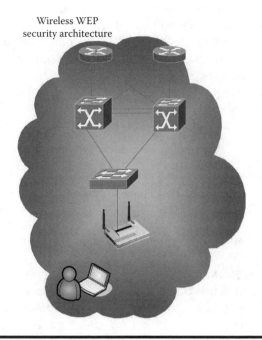

**Figure 15.2   Static WEP wireless architecture.**

these technologies and this security standard. Figure 15.2 details how the network should be set up to support WEP. In this architecture, all the access points should be thought of as an extension of the wired network. Each access point is plugged into a switch similar to how a hub could extend the port count on a wired network.

Some of the major problems with WEP include security and manageability. A large number of attacks have been released against the WEP protocol. These attacks and the amount of risk that they bring are outlined in Chapter 13. The WEP protocol also requires tremendous management efforts. These efforts include the labor involved to distribute the static network keys to all devices that are required to connect to the network.

A solution can be created using WEP; and although it will not be the most secure solution, it is the least expensive hardware-requiring architecture. When talking about cost, WEP is one of the most inexpensive solutions because it is already in all standard 802.11/b/g/a access points. Nothing other than the access points and the client cards are required to support WEP. The support of WEP is already included in the 802.11, 802.11b, 802.11a, and 802.11g standards. The manageability of WEP is very labor intensive, which can change the amount of money needed to support this solution.

Some organizations never take into account the support and operational costs of solutions when they compare them to each other. This after-the-fact cost can equal a large amount. Make sure that the whole gamut of costs for design, install, and support is taken into account when comparing any number of solutions.

Using a WEP solution for small to mid-size companies can prove inexpensive in the short term and costly in the long term. The major cost of WEP is not in the deployment; performing a first-time setup of WEP is simple even in large companies. Tools are available to set up clients with simple executable files. These can be pushed with Microsoft group policy or manually installed by IT staff members.

The real cost of WEP lies in support and ongoing key management. The first time a key is compromised, a full re-key of everyone is needed. This means the cost associated with the security keying of the first deployment of clients will become a reoccurring cost until another security solution is deployed.

Looking at WEP from a risk perspective, it is one of the most risky wireless security standards out there. There are a number of tools available to crack WEP; most of them are detailed in this book. Another downside to WEP is management. As said before, re-keying is a major effort. On top of this, re-keying involves an all-or-nothing approach. In a WEP environment, one cannot take a phased approach to updating the keying information; it all must be done at the same time. This means all clients and access points should be re-keyed at the same time. In a large environment, this can pose a major problem and the outcome could be a decision not to re-key. One of the other issues with WEP is that it is only a computer-based authentication mechanism. This means if someone were to steal a laptop already set up to connect to a WEP-enabled network, that machine would connect no matter who was using it.

## 15.2 VPN

The next architecture is the virtual private network (VPN) approach. In this architecture, one takes all the wireless airspace and treats it as a public network such as the Internet. This will be done at a policy level and at a technical level, segmenting the wireless network and locating it outside the local trusted network. This would mean that all wireless access is very similar to remote access. The wireless clients will have to establish a VPN session with a wireless firewall, gateway, or VPN concentrator on the inside of the network. Before proceeding into this, we outline some of the technologies used in creating a VPN. This will provide the knowledge needed to apply this technology to wireless. Given a good understanding of this, we can then look at two approaches and how they apply to small, medium, and large companies. At the end of this section, we address the commonly missed policy changes that should be made to accommodate a VPN wireless architecture.

## 15.2.1 *Technology Overview*

This subsection discusses virtual private networking. This technology creates encrypted tunnels through public networks. This is done to connect two private networks together. It can be extended to connect multiple private networks into an encrypted extranet. One of the largest examples of this is the automakers' ANX network. This network consists of multiple IPSec VPN gateways, all connecting to each other form a large protected, secure network. VPNs protect confidentiality and integrity with the use of encryption. Many companies have some type of VPN supporting remote offices, remote users, or partner access.

### 15.2.1.1 *IPSec*

The IP Security protocol (IPSec) is an open framework made to give different manufacturers the ability to create secure tunnels. How this is done is through the use of a set of rules that dictates how two devices will communicate with one another and how they will go through the process of negotiating, creating, and tearing down an encrypted tunnel between them. IPSec allows multiple vendors to establish encrypted tunnels without having to have them work together on the development of their products. If their products are properly IPSec compliant, then they should work with other IPSec-compliant products. This has allowed for a major increase in the use of VPN technologies and made the use of VPNs commonplace in most businesses.

Before diving into each part of IPSec, one needs a general understanding of how IPSec works and what each part does. The first portion of IPSec addressed here deals with the creation and exchanging of keys. These keys are used to create the encrypted tunnels used for secure communications through public networks. Subsequently, this section discusses the two main methods of IPSec operation: (1) Authentication Header (AH) and (2) Encapsulating Security Payload (ESP).

### 15.2.1.2 *ISAKMP*

The Internet Security Association and Key Management Protocol (ISAKMP) is a framework provided for IP services to manage security associations

(SA). In this case, the IP service is IPSec. The security concepts of authentication, key management, and security associations are the focus of ISAKMP. The current ISAKMP standard is outlined in RFC 2408 (located on the IETF Web site). One of the goals of ISAKMP was to have a key management protocol that is independent of key exchange. This means that while ISAKMP creates, deletes, and manages security keys, it does not have the ability to exchange the keys with other parties.

### 15.2.1.3 Internet Key Exchange (IKE)

For VPN key exchange, one can use a method called Internet Key Exchange (IKE). The IKE protocol is used to set up the key exchange portion of a VPN. Each VPN can run a number of encryption types consisting of different ciphers with different key lengths. After creating these keys using ISAKMP, it is up to IKE to exchange them between the two parties participating in a VPN tunnel. Creating the keys occurs outside of IKE; only the transferring of keys between two parties is part of IKE. This was one of the goals of the IPSec VPN creators; they wanted one protocol to create the keys and another one to exchange them. This is where ISAKMP and IKE fit in; ISAKMP creates and manages the keys, and IKE actually performs the exchange.

### 15.2.1.4 AH

One uses the IP Authentication Header (AH) portion of IPSec to provide connectionless integrity and data origin authentication for IP datagrams. It was outlined originally in RFC 2402 and then updated in December of 2004 and renamed under the IETF as draft-ietf-ipsec-rfc2402bis-10.txt. This portion of IPSec can be selected for use to provide additional protection on an IPSec data conversation. AH works by encrypting the IP header information and creating an encrypted hash used for integrity checking. Not all the IP header information is used in this hash. This is because some of this data changes as the packet moves across a network. The AH portion of IPSec can be used by itself or in conjunction with ESP to increase the security of a packet protected by IPSec. The Internet Assigned Numbers Authority (IANA) has assigned a Protocol Number of 51 to AH.

### 15.2.1.5 ESP

The IP Encapsulating Security Payload (ESP) is a mode in which IPSec can run; it is outlined in RFC 2406. In September 2004, this was updated, creating an IETF document named draft-ietf-ipsec-esp-v3-09.txt. ESP provides confidentiality, data origin authentication, connectionless integrity, and anti-replay service to IPSec data communications. This is done by encrypting the data portion of a conversation and placing the encrypted data back into the normal data portion of an IP packet. This process can apply encryption to TCP, IP, UDP, and ICMP. To identify the packet as an ESP encrypted packet, an ESP header is inserted after the IP header.

ESP can operate in two modes: (1) tunnel and (2) transport. The difference between these two modes has to do with providing extra confidentiality assurance by encrypting the IP header. In tunnel mode, the real IP address of the client as well as the other party's IP address it wishes to talk to is encrypted and placed into the data section of the packet. For IP communications to work, the two IP addresses of each IPSec gateway are used to communicate with each other. They send encrypted packets to each other on behalf of all participating entities behind them. In a wireless situation, this is usually not used because the IPSec implementation is client to gateway and not gateway to gateway. In transport mode, the IP header information is not encrypted or used with ESP. The Internet Assigned Numbers Authority (IANA) has assigned a Protocol Number of 50 to ESP. This is always present in the ESP header and is used to identify the packet as one carrying ESP encrypted data.

## 15.3 Wireless VPN Architecture Overview

When creating a VPN wireless architecture, one first needs to set up wireless as a non-secured area. This means it needs to be logically separated from the internal network. If the wireless is located outside the internal network, a hacker who compromises the security has nothing of value to connect to and steal. Looking at this architecture in Figure 15.3, one can see that the wireless end devices are outside the internal network, located in a demilitarized zone (DMZ). The internal network is protected from the wireless by a firewall and behind that is a VPN concentrator. This VPN concentrator device is where the wireless end devices terminate their IPSec encrypted tunnels. When an end device wants to connect to the network, it first connects to the wireless without security or with a very limited set of security controls. Once this connection is made to the wireless network, the client then makes a VPN request to the VPN concentrator located inside the protected network. This request is allowed to pass through the firewall to the VPN concentrator. Once the request arrives at the VPN concentrator, an encrypted tunnel is set up, allowing the wireless client to make a secured connection to the internal network. This protects the internal network from any wireless attacks launched and

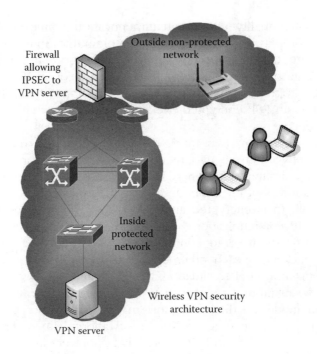

**Figure 15.3   Wireless VPN architecture.**

makes the risk to the wireless device the same as Internet-connected devices.

When discussing the wireless VPN architecture, there are two main approaches for the VPN concentrator; one can locate them either remotely or locally. In locating them locally, one VPN concentrator would be at each site where there is a wireless network. In a remotely placed VPN solution, a single VPN concentrator would be located at a headquarter location or high-speed data center. This single location would service all the sites. When using each of these approaches, there are many advantages and disadvantages associated with them. The next two paragraphs discuss the pros and cons of each approach.

The first approach is to place the VPN concentrators locally. In this scenario, one has a VPN concentrator sitting locally at each site. All encrypted wireless traffic is terminating into this device, allowing it to be decrypted and sent into the network. This solution provides quicker response and better availability than the centrally located solution. In the event of a wide area network (WAN) failure, this solution will still operate — unlike the centrally located solution. In addition, this solution has a more secure architecture because the traffic does not have to cross over the internal network for authentication and encryption. Some of the cons to this solution are the higher cost of electronics, as well as the added cost to deploy and administer multiple VPN devices at different locations.

Using the second approach, one would place the VPN concentrators remotely. This creates a slightly different story compared to having the concentrators locally placed. In this scenario, one would have many sites using the same VPN concentrator for authentication and encryption. When the wireless end devices want to connect to the internal network, their request data must travel over the WAN to the location of the centrally placed VPN concentrator. This means that there is a security risk for this data to mix with other confidential data in transit. Another big issue with this architecture is that a loss of the WAN means a loss of the wireless. In some cases, the benefit of low overhead and administration costs may outweigh the security risk and available risks. On the plus side of this option, one sees a large up-front as well as TCO (total cost of ownership) cost saving compared to the first approach.

Looking at the VPN architecture as a whole and comparing it to other solutions discussed in this chapter, one can discern the pros and cons. First, the architecture has been proven secure with wide usage across the Internet. Today, many people use VPN technology across the Internet, which is much less secure than any wireless network. In addition, the VPN architecture can support multiple authentication methods and integrates well with multiple factor systems.

One of the major cons related to a wireless VPN architecture is the cost of the solution compared with other available options. The next major con related to this particular architecture is the lack of consistent roaming from one site to another. A major security issue is the access points themselves. They are out in the DMZ, making them very vulnerable to attacks. A hacker could attack the access points and cause denial of service. A hacker could also put up his own access point in hopes of stealing VPN credentials. Another issue with hackers is the likeliness that they will connect to the network regularly, although they will not stay due to the other layer of security preventing them from connecting to the internal network. In addition, this approach will most likely get the network listed on war driving Web sites because its existence is easily detected.

## 15.4 VPN Policy Aspect

Using this VPN architecture requires more than just some technical challenges. Some policy changes must be made to account for the new wireless network. Because the wireless is outside the local network and not directly connected to the Internet, it needs to be addressed in the security policy separately and correctly. Using some of the ideas that have protected companies from threats on the Internet can easily carry over to wireless.

A high-level policy should address wireless within the same risk factor as the Internet. If a more detailed examination is required, someone could reduce the risk factor of wireless compared to the Internet because the Internet has a higher number of malicious users than a wireless network. In looking at the easiest way to perform this type of analysis, taking the approach of treating the non-secured wireless network similar to the Internet holds merit. Anyone could, would, and can get onto the network and possibly do harm. Addressing both the Internet and wireless with this mentality will help identify what current controls are in place to protect the local network from the Internet and that can be carried over to the wireless to accomplish the same risk reduction. What this means is that if a company has decided that two-factor authentication is required for Internet VPN devices, then the same can be applied to wireless. If the same safeguards that apply to the corporate remote access are applied to wireless access, the risk of exposure is reduced. Any policies that apply to connecting to the corporate network from the Internet may also apply to the wireless. Looking at each of these policies will save time compared to creating new policies to address similar access methods.

## 15.5  Wireless Gateway Systems

The next architecture is the wireless gateway or firewall. In this architecture (Figure 15.4), the wireless network is segmented from the wired network in a similar way as in the VPN architecture. This means the wireless access points and their clients are on a DMZ outside the corporate firewall. With this architecture, the firewall or gateway allows or denies the wireless end devices' access. This access can be set to different resources beyond it. Some of these devices have the ability to select groups based on user credentials and allow or deny the client to different portions of the internal network. For example, a group could be created only allowing the wireless end devices to access the Internet or maybe a single server sitting deep inside the internal network.

Some of the different vendors offering a gateway solution include:

- *Air Fortress:* http://www.fortresstech.com
- *BlueSocket:* www.bluesocket.com
- *Vernier:* www.verniernetworks.com

These three gateway manufacturers have many other competitors not named here. The selection of these three manufacturers and their products

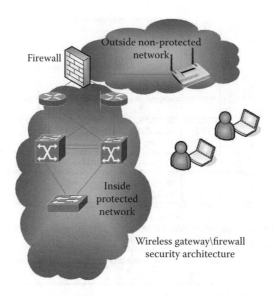

**Figure 15.4   Wireless gateway.**

is only so that one becomes familiar with the various ways each manu-facturer uses technology to accomplish the goal of protecting and authen-ticating wireless users. In progressing through this chapter, one will see that each of the three devices was chosen because each has various features that are unlike the other two. Some of these features can be found on other products outside the three listed here or may be unique to that particular product. These products are listed here more to inform one of the available methods of using a wireless gateway and how that influences the architecture.

In progressing through this section, one will begin to see that one product may have a feature that one needs. For example, maybe one needs a clientless solution or a military-grade solution. Each one could have an impact on the architecture of a gateway solution. Understanding up-front what requirements are needed will ensure that time and effort are not wasted evaluating solutions that will never materialize.

When one looks at using wireless gateways or firewalls, one must contrast the pros and cons of any architecture. Looking at the pros, one first notices that guest sign-on is supported by almost all wireless gateways or firewalls in this category. One also notices support for access control lists that are based on a number of advanced objects. These objects can be accounts, devices, or network segments. Having flexibility in a security product is always an added bonus. During the life cycle of any security appliance, user requirements are often pushing device features that should be limited to only a select few. Being able to accommodate these select

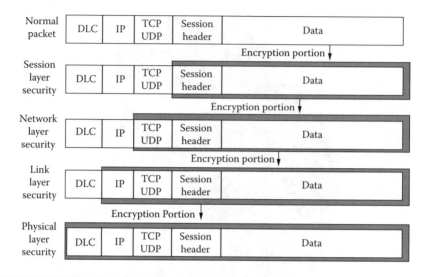

**Figure 15.5  Encryption based packet layers.**

few without giving everyone else the same level of access is what makes these devices shine.

Each device has different features and functions as to how they operate. Some devices, such as Air Fortress, can encrypt communications at the data link or MAC layer. This is beyond what any VPN device can do because it needs the MAC and IP layer to send and receive packets. A VPN device can mask the IP layer using only a gateway-to-gateway communication structure, although this is very unlikely in reference to a wireless VPN solution. To get a good idea of how this works, look at Figure 15.5. This figure shows the various layers of network communications, detailing how devices such as the Air Fortress can encrypt data communications at layer 2. When two devices are talking, all one can see is the sender's and receiver's MAC address; all other communication is encrypted.

In discussing the encryption of gateway devices Air Fortress comes up as one of the strongest encryption providers with support for AES. From Chapter 12 one learned that AES is the current government standard for encryption. Air Fortress was first to receive the government's FIPS 140-1 (June 2002) and performed recertification in February 2004 to meet FIPS 140-2. This makes this product a very good fit for high-security environments like the U.S. Army. Currently, Air Fortress products are protecting more than 12,000 U.S. Army networks.

The Air Fortress products have the ability to perform three-factor authentication as well. This is done based on the following three factors:

1. *Access ID.* The access ID (or network authentication) is a password located inside the software on each client device. This password is only known to a security administrator who configures the end device and not to the end user. This password allows encryption to take place in order to perform authentication with the other two factors.

2. *Device ID.* The device ID (or device authentication) is a key value created when the Air Fortress software client is loaded on an end device. This device ID allows an administrator to prevent a device from entering a network if it was lost or stolen by simply rejecting its device ID. This allows another option, that of only allowing device authentication and not the next factor of user ID to take place. This would be common for low-level devices that do not support user-based authentication. A good example of this would be a handheld scanner.

3. *User ID.* The user ID (or user authentication) fits the "something you know" factor. The user ID is a password required by the system to identify the user of a device. The user ID supports back-end user credential and identification systems such as RADIUS, NT Domain, Active Directory, and LDAP. This user ID can be used with or without the device ID. When using the user ID with the device ID, a deep level of granularity can be achieved. This gives the administrator the ability to limit connections to single devices by user or to allow the user to use multiple devices. For example, one can set up three devices that are the only devices allowed to be accessed by a selected person.

In other locations such as schools, we have seen significant usage of the BlueSocket gateway product. This is largely due to the important feature of requiring no client or added software to support its use. This has helped in environments such as schools where there are many people and devices that frequently change semester to semester. This feature is only available on a few gateway products out on the market. This is primarily done with the use of an SSL-protected user authentication. This authentication takes place on a Web page controlled by the gateway. This solution may be a feature that can lower the cost and administration by removing the need for a client, although it will increase risk compared to solutions that have static devices defined by software and configuration. To get a higher level of security as well as to utilize other advanced security methods, a client is needed.

One of the areas that BlueSocket expanded into was rogue access point detection. BlueSocket devices now have sensors that can talk to a server, indicating the presence of a rogue access point or someone trying to launch a number of known wireless attacks. One new move in this area made by BlueSocket is to give the sensors some intelligence. This allows them to process the airwaves and only send traffic that it thinks might be a threat to an upstream device for further processing. This allows for bandwidth savings and central reporting and monitoring capability.

Another area that BlueSocket products have expanded into is layer 3 roaming. BlueSocket gateways allow a user to roam seamlessly from one subnet to another without any loss of communications. This is something that will become big value in the near future as we start to see more and more communication devices traveling over IP-enabled networks.

The final manufacturer of a gateway system mentioned here is Vernier Networks. Its gateway has a very different approach compared to the other two. Vernier's gateway system can be used for wired and wireless clients, providing authentication and policy management to all clients. This device has taken a full security device approach to the gateway architecture. It allows for worm and virus detection as well as containment. It allows for easy access granting or detail for sections of the network. Using Vernier's gateway for wireless allows for a similar clientless connection like BlueSocket or a VPN approach leveraging IPSec. This product is more gauged for a complete security gateway, providing malicious code detection and containment on top of basic access control and authentication.

Looking at some of the disadvantages to this architecture as a whole, one notices that these devices do nothing to protect transmission over the airwaves. All the security features of these devices take place only once the data has crossed through the product itself. This means that all the wireless threats can be exploited over the airwaves until they pass through the gateway device. The only slight mitigation is seen with the advent of link layer encryption.

One of the advantages of this is in the event that the wireless becomes compromised, getting into the network still requires defeating the security of an appliance. Another drawback to this architecture is one that also plagued the VPN solution — the need for device hardware and support. Each device should be purchased and supported. This is always a sore point because it involves many dollars in a large situation.

## 15.6 802.1x

The 802.1x architecture involves another approach different from the ones discussed previously. This architecture is the general direction of the IEEE 802.11i standard. A number of companies have taken this direction to lay

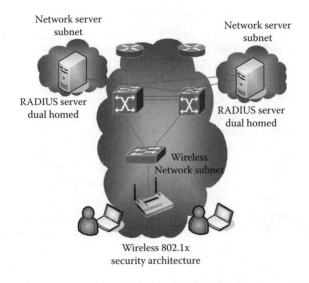

Wireless 802.1x
security architecture

**Figure 15.6   802.1x Wireless security architecture.**

the groundwork for the 802.11i standard. Currently, the 802.11i standard is out, although many products are just starting to utilize it. This is the primary reason why an 802.11i architecture approach is not included in this chapter. All the pieces that make up 802.11i are located in the 802.1x architecture. This 802.1x architecture involves keeping wireless networks safe using an EAP-encrypted channel to send authentication traffic into the network. Once validated, one is granted access to the network. Until this validation takes place, the only traffic allowed to pass into the network is the EAP authentication traffic.

Looking at the architecture detailed in Figure 15.6, one can see that no longer is the wireless considered hostile and in need of separation from the rest of the network. Now the wireless is located on the inside of the network. This is due to the prevention of traffic flowing through the access point into the network. With the 802.1x standard, the only traffic one can pass without authentication is authentication traffic. To support this architecture, one needs the three key pieces that make up the 802.1x standard. The supplicant is needed and is present as a wireless client. The authenticator is needed and is made up, in this case, of the access points. The third piece needed is one that is added from the other solutions, the authentication server. In this case and most other 802.1x cases, a RADIUS server is used to fulfill this piece.

Now that each of the pieces has been identified, one can see the solution in action. When a client connects to the network, the access point will ask it to provide authentication via EAP. Depending on which EAP type is used, different authentication methods inside each EAP type take place. No matter what EAP type is used, a successful authentication is signaled to the access point by an EAP success message. Once this message is relayed from the authentication server to the access point, the access point will let the supplicant or wireless client onto the network.

We are going to look at the three main vendors in the 802.1x authentication environment. Cisco, Microsoft and Funk have had products out for some time now, products that fit well into this architecture. Looking at each of them, one can see the benefits and features that each one brings. All three bring the RADIUS server piece into this architecture. This is because the other two pieces (authenticator and supplicant) are already defined as a wireless client and an access point. Depending on what EAP type is used or what features are needed, each of these RADIUS servers may or may not fit into the architecture. This is where clear requirements are needed up front. Cost can always play a big role and as one will see, certain servers have different methods of costing associated with them.

Our first 802.1x solution is based on Cisco's access control server (ACS). This is an authentication server that can have account credentials locally or can point back to a common directory structure such as Active Directory or Novell's Network Directory Service. The ACS server allows for the usage of MS PEAP client, which ships with Windows XP service pack 1 and is available via a free downloadable patch for Windows 2000. The ACS server also supports Cisco's own LEAP and EAP-FAST client and Funk Odyssey PEAP and LEAP client. This RADIUS server allows for a large amount of EAP types, including most of the well-known types. The cost of this solution is based on a per-box cost. This means X amount of boxes will cost X amount of dollars. This may seem a little obvious, although as we progress through this section we will see other ways of costing a RADIUS server. This will make us appreciate the up-front signal cost per box model.

The next architecture is based on the Funk Odyssey RADIUS server. This server has the same ability to create accounts local to itself as well as to point back to another directory structure. It also supports the same client base that Cisco supports with an added Funk EAP-TTLS client available. One thing to note about the Funk system is its added support for older devices such as DOS terminals. Funk has a client for DOS to aid in getting these older devices onto the network securely. This ability to allow older clients onto the network has fulfilled a requirement that no other RADIUS servers have. Funk costing is performed on a per-client license. This is very different from a per-box cost. Not only do you have to pay for the RADIUS server hardware, but also you have to pay a licensing fee for every client running on the network.

The last 802.1x architecture is present in most companies already. This has made it one of the best financial choices of the three. The MS-RADIUS server is a function built into Windows 2000 or 2003 Server. Microsoft supports the RADIUS standard on both platforms, although it highly recommends using Windows 2003 over Windows 2000. In the MS solution there are some drawbacks, such as no support for any other authentication database other than Active Directory. This is usually not too big an issue, seeing how the fact that this solution would be considered would have to do with the already-existing Active Directory (AD) architecture. If a customer did not have a Windows network with AD, the cost saving benefits would not be there. This would leave this solution with a slightly low cost, although it would also bring the most limited set of features and modes available. One also notes that there are no patches for older devices. Windows 2000 Professional, XP, Pocket PC 2003, and all CE.NET devices are supported. This means no Windows 98 or DOS terminals under this solution can be used. This solution also does not support Cisco LEAP or EAP-TTLS. The only EAP types supported are EAP-MD5, PEAP, and EAP-TLS.

Some of the advantages of this solution have to do with the cost and support savings of not having to separate and keep securely separated wired and wireless networks. Another advantage has to do with the ease of upgrading to the 802.11i standard once it becomes widely available in products. Moving from one of the other architectures to 802.11i would require massive network architecture changes. With the 802.1x solution, moving from it to 802.11i only means having to change the access point configuration and possibly the client cards to support AES.

Some of the disadvantages of this solution are similar to the other network types. War drivers can still see the wireless network. This is not as bad as the other solutions where almost all traffic is seen in the clear. In the 802.1x solution, all management traffic is seen in cleartext. Being able to see this traffic means that hackers can perform denial-of-service and man-in-the-middle attacks. A number of attacks that can be done on an 802.1x network are outlined in Chapter 14. Those relevant to this section have to do with older RADIUS servers. When 802.1x and RADIUS were first applied to wireless, a number of issues surrounding dynamic key creation and exchange were not defined in the 802.1x standard. This meant that a number of older RADIUS servers always served up the same WEP key for everyone. This is nonexistent in newer versions of the three RADIUS servers listed here. This means, however, that older versions of these already deployed servers are not performing this action today. Another issue has to do with the non-authentication of the network to the user. This issue can be mitigated with the use of certain EAP types that authenticate the network to the user.

## 15.7  Comparing Wireless Security Architectures

This section compares each of the architectures discussed above. Performing these evaluations is a step that should be done for each unique business environment. This evaluation is done at a very high level, pointing out some of the major issues, risks, and hidden costs. This will help to narrow down the choices that would apply to a certain situation. This can help reduce the time and effort needed by eliminating one or two architectures, thus allowing a team to focus on one or two of the viable solutions instead of evaluating a large number of solutions.

This section also helps one understand the risks involved with each approach. If someone comes into an existing network, having an understanding of the general architecture approach taken will help. This section also gives an auditor an understanding of how to weigh the risks in each of the architectures with respect to security.

### 15.7.1  WEP Architecture

Looking at the WEP architecture as applied to a small network, one can see that the relatively low cost of initial deployment is a plus when compared to other approaches. As stated previously, WEP is initially the most cost-effective solution until one adds in the extra support dollars needed to manage it. One can also see that the major burden of management may not be as daunting a task in comparison to larger networks. If there are not as many access points and clients to manage, then the overall cost may be more competitive when compared to other, more easily managed solutions. The final point about applying this solution to small network has to do with risk. Depending on if this network is a sole network, only existing at a single location, the whole risk that WEP brings may not even be that applicable to the business. For example, if the business is a manufacturer of cardboard boxes and its network only consists of a few wireless scanners, ten desktops, and one server, this weak wireless security solution may be a low-cost solution with some added risk. Performing a risk assessment and understanding what electronic assets are most critical to the survival of a business will help to weigh the cost versus security in any security architecture.

Carrying this design over to a mid-size network, one starts to see that the overall cost is large enough to make the added expense of a layered security solution a small cost in comparison to the overall project. In a mid-size company, the number of access points and clients will be high, thus increasing the cost of management. Using WEP in this case is the best solution when comparing up-front costs, although it could be the most expensive total cost of ownership (TCO) solution for wireless security.

Looking at the risk of WEP, one should know by now that it is the most risky solution discussed herein.

Using WEP in a large network makes the issues concerning management even more costly. Also, in large environments, the cost of an advanced security solution would constitute a very small piece of the overall cost of deploying wireless in this scenario. So, by far, WEP should never be one of the best solutions for a large network, even if the risk factor has nothing to do with the decision-making process.

## 15.7.2 Wireless VPN Architecture

When looking at this solution in terms of the three scenarios, one sees that in a small network with fewer than ten users, the cost of a VPN device may be more than the total project cost for the four or five access points needed. This makes the added security a major expense. The manageability of the solution is low compared to WEP, although each user will need software and some type of configuration performed. This is a small cost in this environment. Next, looking at the risk, this solution is acceptable for a number of vertical markets and, depending on what authentication and encryption types are used on the VPN, it could be applicable for high-security environments.

In a mid- as well as large-size network with multiple locations, a choice must be made as to placing the concentrators locally or remotely. The benefits, costs, and risks were discussed previously. Once a choice is made as to the placement of the VPN device or devices, a closer look at the cost and support is needed: how much it will cost to support, added training for personnel to learn about VPN technologies, and added cost for help-desk training to help users with their VPN client. This could actually prove to be the opposite if the company already has a VPN solution for remote users. If this were the case, then the support, training, and administration costs would be significantly reduced, because the support personnel already know how to support the device. In a large network, this most likely would be the case.

One of the biggest management costs beyond general support would come over time and would not be seen or realized up-front when the project first completes. This cost would be associated with ensuring that the wireless network stays confined within a DMZ. Some people will have issues with the VPN access that they have been granted and will perform changes to help circumvent the VPN altogether. This could be seen in added access points located inside the network, either rogue or company owned. This could also be seen in moving an access point logically from the DMZ to the internal network. Those who appear to have issues with using their VPN connections might choose to set up their own wireless instead of working out their VPN client issues. This means the tracking and location of rogue access points now becomes an ongoing cost. Rogue access point detection should always be performed, although this scenario is more likely than others to have rogue access points placed by employees.

Looking at this as a whole in a mid- or large-size network, the total cost of the security architecture becomes a small percentage of the total cost of the project, making it a more applicable choice. The risk involved in this solution is better than in most other solutions. Using the VPN architecture can allow for a wide variety of authentication methods, thus increasing the overall security. Some of these methods are tokens, certificates, or just plain old passwords.

### 15.7.3 Wireless Gateway or Firewall Architecture

Most of what was discussed regarding the VPN architecture fits into this approach. Using this in a small office could be costly. Depending on if this architecture utilized a clientless solution or not could make it less of an administration effort compared to the VPN and WEP approaches. When a higher level of security is needed and client software is required, the support cost of this solution starts to become close to that of the VPN solution and not too far behind the WEP solution. This solution in a small organization is based on two key factors: cost and security. The cost of the hardware will be present regardless of what security method is chosen. The other major factor in this solution is the choice of advanced security or average security. Depending on the level of security, the cost of support will increase.

Using this approach in a mid- to large-size network, some of the same equipment costs as the VPN architecture present themselves. Placement of the hardware and security level required makes the cost of hardware and support increase. Using clientless solutions reduces this cost, although an increase in risk follows when compared to a client-based solution. Another big issue that affects cost in larger networks is the support costs of finding rogue access points and making sure all access points are located in the DMZ. In a large network where there are many geographically dispersed locations, finding and preventing rogue access points from appearing on the internal network can become a large, costly task.

### 15.7.4 Wireless 802.1x Architecture

This solution can be applied easily to a small network if the organization decides to go with a Windows RADIUS server and has an existing Windows network. If these two requirements are already decided, then adding the wireless 802.1x architecture only involves configuration changes on a Windows server and the access points. If this is not the case, and the small site needs another RADIUS server outside the one that is included with Microsoft, then the cost of a new server can become a concern. It is always possible to centrally locate a RADIUS server and save some money in relation to the cost of the hardware and support. There are many pros and cons with regard to local or remote placement of any access granting hardware.

So if one were to go with the 802.1x solution and needed support for older devices, then using a Funk RADIUS server may prove the best option. In this case, there are some major costs involved. A server must be procured, installed, and supported. Software, including the operating system and the Funk software, must purchased as well. The last thing that must be purchased is a license for every wireless client. This license must be maintained from year to year.

Looking at a mid- to large-size environment, one can start to see where an MS RADIUS solution can fit well. Most mid- to large-size networks run MS products and have domain controllers that process network authentication. Leveraging these to perform RADIUS is a small task with little resource load to the device. DHCP and DNS are much more resource-intensive services than RADIUS. If one were to use the existing domain controllers, this solution would be the least expensive compared to all the other solutions discussed. This is because there is no added hardware or additional devices to support. Any support already existing on the domain controller can easily be carried over to support the added RADIUS service. Now, if the MS RADIUS server will not fit due to whatever reason, one must account for the cost and added support of an additional device.

## 15.8 Chapter 15 Review Questions

1. What is the most inexpensive architecture to deploy, based solely on hardware?
   a. WEP
   b. VPN
   c. Wireless firewall or gateway
   d. 802.1x

2. What architecture is most risky in terms of security?
   a. WEP
   b. VPN
   c. Wireless firewall or gateway
   d. 802.1x

3. In a VPN, what protocol creates the keys?
   a. IPSec
   b. IKE
   c. ISAKMP
   d. AH
   e. ESP

4. When considering a VPN architecture, what two main options do you have to choose from?
   a. Locally placing devices
   b. Remotely placing devices
   c. AH
   d. ESP

5. When looking at the 802.1x architecture, what RADIUS server has the cost of licensing along with the cost of the server?
   a. Cisco
   b. Microsoft
   c. Funk
   d. BlueSocket

6. When looking at a wireless gateway, what vendor would be the best to use if a clientless solution is needed?
   a. Air Fortress
   b. BlueSocket
   c. Vernier
   d. Cisco

7. In a VPN, what protocol exchanges the keys?
   a. IPSec
   b. IKE
   c. ISAKMP
   d. AH
   e. ESP

8. When looking at a wireless gateway, what vendor would be the best to use if a high-security, military-grade solution is needed?
   a. Air Fortress
   b. BlueSocket
   c. Vernier
   d. Cisco

9. What cost savings can be seen when using a Microsoft RADIUS solution if the client has an existing MS network in place?
   a. Hardware
   b. Software
   c. Installation
   d. Configuration

10. Name one of the largest hidden costs associated with the VPN architecture.
    a. Cost to support the hardware
    b. Cost to support the client
    c. Cost to keep the wireless in a DMZ
    d. Cost to keep the hardware running

11. Which gateway product provides services for worm detection?
    a. Air Fortress
    b. BlueSocket
    c. Vernier
    d. Cisco

12. Which IPSec mode is most common for wireless?
    a. AH
    b. ESP
    c. IKE
    d. IPSec Mode 2

13. Which two architectures require moving the wireless into a DMZ?
    a. WEP
    b. VPN
    c. Wireless firewall or gateway
    d. 802.1x

14. Which architecture is the most difficult to change after a single key is compromised?
    a. WEP
    b. VPN
    c. Wireless firewall or gateway
    d. 802.1x

# Chapter 16

# Wireless Tools

Wireless networks pose a threat to all who use them. Wireless networks lack the safety associated with having one's communications securely transmitted inside physical cabling. With wireless communications, the air is the transmission medium and that medium is accessible to anyone who wants to listen. This has led to many tools, which were developed to aid the curious in eavesdropping on other people's wireless communications. These tools are most often created for network management or trouble-shooting and include some powerful capabilities. Often, these same tools have a very dark side if used for that particular purpose. Some of the tools were made for the sole purpose of attacking wireless networks. People often ask why these tools are made and why legislation has not been put into place to stop them. The answer to this question is one that is often debated. The most common answer to this question is that these tools educate and make the public aware of what an attacker can do and how easily they can do it. Most hacking tools stem from a vulnerability or flaw that already exists. All that the tool writers do is make it easy for people to perform the same type of exploit without having the knowledge level required to find it in the first place. Face it; if one can download a program, run it successfully, and defeat one's security, then one might want to start thinking twice about who else might want to try the same steps. Many people have debated the value of releasing tools that circumvent or break security methods. No matter what they say, most all tools come from an already-existent security problem, not a new one created by the tool.

The tools used on wireless networks have a few main functions, to include scanning, sniffing, cracking, and causing a denial-of-service attack. Numerous tools currently on the market perform the exact same functions as other available tools. To obtain a full understanding of what tools are currently available, one should be aware of the four primary purposes of wireless tools. This section categorizes some of the existing tools in terms of their functions and features.

# 16.1 Scanning Tools

Scanning tools are easy-to-use sniffers that capture network data and provide a GUI (graphical user interface) for the user to view information relating to wireless network identification. The scanners make it easy for someone to see the type and settings of a wireless network. This is very useful for someone looking for networks in a war drive. The main function of a wireless scanner is to find wireless identification information. Some of the information that these scanners will find includes channel, security type, MAC address, access point vendor, data rate, signal strength, noise level, signal-to-noise ratio, and GPS location. Other than GPS, all of the above information can be found with a wireless sniffer. Scanning tools allow users to see very easily what networks are in their area and what their settings are. Although sniffers can find this information, these scanners do a much better job of presenting the relevant information to the user. These kinds of wireless tools are what are commonly used for wireless network identification. Most war drivers use these wireless scanners to find networks and then move to a sniffer to get information once they have selected a target.

## 16.1.1 Network Stumbler

On May 5, 2001, Marius Milner released Network Stumbler to the world. This wireless network scanner quickly became one of the most used and liked wireless scanners available. The first version was very limited in its ability to work with wireless adapters and operating systems. Since then, a number of improvements and version numbers have been created. The most current release is version 4, which was released on April 21, 2004. This version supports many wireless cards and can run across most Microsoft platforms. The Network Stumbler software is available from the Web site located at www.netstumbler.com.

Network Stumbler has all the abilities and features outlined in the general section on wireless scanners. It can find the existence of a wireless network listing, its channel, and if WEP is used or not. It has the ability to associate common MAC addresses, identifying the access point manufacturer to the user. The tool can find the SSID information used to identify the wireless network from others in a given area. The program also has the ability to map signal strength and noise levels. This feature has made Network Stumbler a troubleshooting tool as well as a wireless scanner. Using Network Stumbler as a troubleshooting tool is an easy way to have a great troubleshooting tool — especially because it is free. One of the great benefits of Network Stumbler in relation to wireless network troubleshooting is the ability to see multiple access points and their corresponding signal and noise levels. This is very useful in an environment where there might be too much coverage provided. In this case, a wireless client will hop from access point to access point so fast that it degrades the overall throughput. This degradation comes from the management messages needed to roam from one access point to another. With Network Stumbler, a troubleshooter has the ability to see multiple access points and determine that too many of them have the same signal levels causing this type of issue.

Getting into how the Network Stumbler program operates and what it looks like, one first needs to download the software from the official Network Stumbler Web site. Once this software is downloaded, it can be installed on a wide variety of Microsoft operating systems. Once the software is downloaded, unzip it and click the executable. This will start the installer; just click the defaults and the product will install.

In using Network Stumbler for the first time, one will see a user interface similar to the one in Figure 16.1. From this interface, one can perform most of the gathering and evaluation of wireless networks within a given area.

The menus across the top of the screen along with their functions are listed as follows:

- *File menu.* The file menu allows one to save the networks found with their settings. This means one can later evaluate a particular location scanned. The File menu also allows one to open saved scans and perform a merge operation to combine multiple scans into a single scan. One of the other good features available in this menu is the ability to export the scans into Excel, a mapping program called Stumb Verter, or other types of mapping software.
- *Edit menu.* As with most Edit menus, this is where the copy, cut, select all, delete, and paste commands are located.
- *View menu.* The View menu allows one to select how the viewable area will be displayed. One can increase the font size or change

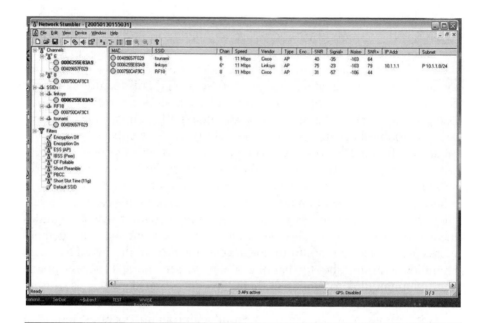

**Figure 16.1    Network Stumbler screen shot.**

the arrangement of scanned networks. The Options tab located inside this menu provides many of the configuration features of this tool.

■ *Options menu.* This allows one to speed up or slow down the scanning speed, display the format of time, change GPS options, and enable or disable scripting.

■ *Device menu.* This menu allows one to select the network adapter one wants to use for scanning. Network Stumbler will most likely select the correct adapter automatically, although it does give one the option to force the use of another adapter if wanted.

■ *Window menu.* The Window menu provides one with more options, similar to the View menu. It lets one adjust what the interface looks like and allows one to use multiple windows to compare multiple scans. Being able to start a new window, for example, when one is in a different area that one wants to scan, can be done from this menu.

■ *Help menu.* The Help menu brings up the help files and the "about program" display. This is where one can get further help with any issue one might be facing. If one cannot find the information one is looking for, try the forum on the official Web site for additional or advanced help.

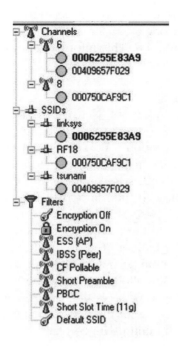

**Figure 16.2    Network Stumbler screen shot filters sections.**

Now take a look at how to use the user interface. On the left-hand side are the two top categories: SSID and Channels. These basic filters are displayed by default. To get more granular, one needs to open the Filters tab. This allows one to sort the scanned networks by a number of setting parameters. The parameters are listed below, along with an explanation of how they work; all the filter sections are shown in Figure 16.2.

- *Encryption On.* This means that the networks have some form of encryption enabled.
- *Encryption Off.* This means the access points are wide open and no encryption is enabled.
- *ESS.* This means the device is operating in infrastructure mode.
- *IBSS.* This means the device is operating in ad hoc mode.
- *Default SSID.* This is a new field recently added to specify if the current SSID is a default SSID created by an access point manufacturer.

In addition to these many features of Network Stumbler, there is also a feature that makes this tool a must-have for war drivers and wireless hackers. This feature is the ability to perform GPS location. The GPS integration allows Network Stumbler to map GPS coordinates to access points found by the program. This information is then used to populate a mapping program to display the locations of access points to the user.

When discussing the use of access point information and mapping programs, another program comes to mind. This program was released to allow for easy integration of Network Stumbler data to Microsoft MapPoint software. This program is called StumbVerter. The program provides the ability to import Network Stumbler's summary files into Microsoft's MapPoint 2004 maps. After someone finds the wireless networks with Network Stumbler and records the GPS information related to each access point, he can then import this information into Microsoft MapPoint with the use of the Stumb Verter program. The logged access points will be listed inside the mapping program as MapPoint pushpins. These pushpins allow for balloons to pop up after clicking on them. These balloons contain information such as access point MAC address, its signal strength, and SSID. These balloons can also be used to write down any other information. Because of tools like this, many people have started to create a large map listing all the access points people have detected by war driving across the entire United States.

### 16.1.2 MiniStumbler

The MiniStumbler program is a tool for wireless scanning that can be loaded on mobile units such as PDA devices. The same person who created the Network Stumbler program for desktops also created this version for handheld PDA-type devices. At first, the version of this program followed its own version advancement timeline. The newest version of this program created an equal version level between Network Stumbler and MiniStumbler. The program operates the same as the desktop version and has the same functionality and features. The supported operating systems include:

- HPC2000
- PocketPC 3.0
- PocketPC 2002
- Windows Mobile 2003

This program is used as an easier-to-carry wireless scanner. This allows users to move freely without having to lug around a heavy laptop. We have also seen this tool used for war driving and war walking. With today's PDA devices and smart phones, more and more people have the ability to war drive as they travel down the street performing their daily routine.

### 16.1.3 Wellenreiter

Another tool used for scanning wireless networks is Wellenreiter. This is a popular wireless network discovery and auditing tool for the Prism2, Lucent, and Cisco wireless client cards. It is called one of the easiest Linux scanning tools available. This tool is very similar to the Network Stumbler tool. The real difference is that Wellenreiter supports the Linux operating system, whereas Network Stumbler supports Microsoft operating systems. Wellenreiter does not have as many features as Network Stumbler, such as GPS integration. This tool can find wireless networks and list their channels as well as their MAC addresses. It can also list the SSID of the networks that it picks up. A need for this tool was realized when the first cracking tools for WEP emerged. These tools only came out for Linux, and not Microsoft. This meant that when war drivers went out, they had to switch between Microsoft and Linux during a war drive. This became an issue, so a Linux version of Network Stumbler was created. This version really has no ties to Network Stumbler, although they operate and function the same. Now Wellenreiter has come out with a new version that can port like MiniStumbler to PDA devices. This was due to a change in the code from Perl to C++. Using C++ gave Wellenreiter the ability to port its software to Zaurus, Ipaq, and other PDA devices.

Figure 16.3 shows a screen shot of Wellenreiter. From this one can see that it is laid out similar to network Stumbler, with some differences in the GUI layout. For example, Wellenreiter does not have the filter section easily available on the right-hand side. It does filter on the right-hand side but only by channel. The newer version of Wellenreiter, called Wellenreiter II, has a better-looking GUI. Both versions have one unique feature in the fact that each puts a logo for a laptop or access point on the devices that its scan detects.

### 16.1.4 Wavemon

This software package, like WaveStumbler, works with Linux and provides access point identification to the user. This software suite allows for graphical mapping of wireless signal levels to hone in on the location of a wireless access point or to perform advanced troubleshooting. Wavemon is an ncurses-based monitor for wireless devices supporting the Lucent Orinoco chipset. In Figure 16.4 one can see Wavemon in action, detailing the signal levels of a Linksys access point.

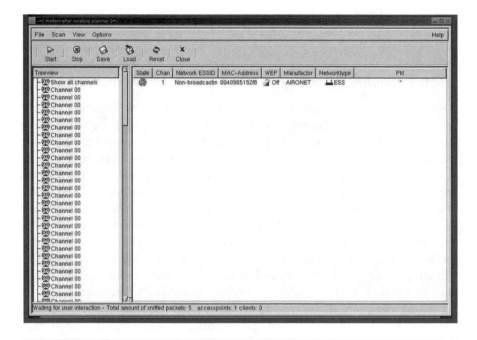

Figure 16.3 Wellenreiter screen shot.

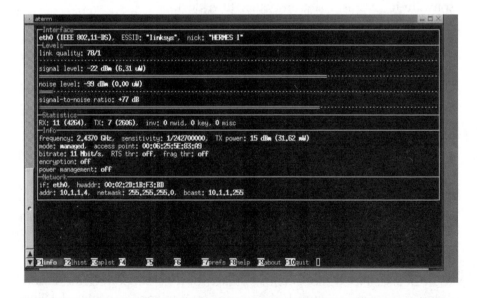

Figure 16.4 Wavemon screen shot.

## 16.2 Sniffing Tools

Sniffing tools allow the user to view packet data on a wireless network. This gives users the ability to see network traffic. This traffic can be seen in real-time based on the type of sniffing tool used. There are many wireless sniffers on the market today. They range in price from high-end, enterprise-type sniffers that cost thousands of dollars to free open source sniffers.

Sniffers are a great tool and can help tremendously in troubleshooting networks; however, in the wrong hands, a sniffer can be used for a variety of malicious purposes. If an attacker uses a sniffer, he can see lots of information about a wireless network. He can see the presence of a wireless network and its SSID, channel, power, and data rate information. With this information, an attacker can connect to an insecure network or understand what type of encryption a network might have. In addition, the sniffer can find IP-related information, which can help an attacker with network mapping functions. This can be expounded by evaluating high traffic talkers based on source and destination. This information can help an attacker find a file server or a critical application based on the amount of traffic sent to it. Most commonly, an attacker will use a packet sniffer to start to mount a man-in-the-middle attack to circumvent any existing security measures. Once this is done, the attacker can get to the next step of performing an exploit and seeing the data unencrypted.

Network sniffers can also find other interesting information relating to weak security methods of older, widely used network protocols. A number of existing protocols have cleartext authentication set as default. Telnet is a good example; if an attacker had a sniffer set up, he could filter for this type of traffic and just capture it. When a network administrator tries to connect to a piece of networking equipment over the wireless, she will be prompted to enter her username and password. Once the authentication takes place, the sniffer will see the cleartext authentication and record the username and password. E-mail is another problem case related to cleartext authentication. POP3 e-mail retrieval is also authenticated in cleartext. This becomes an issue for mobile workers pulling e-mail through a wireless network in a coffee shop or other hotspot. When they request their POP3 e-mail, they must authenticate in cleartext to receive it.

### 16.2.1 AiroPeek

WildPackets created a wireless sniffer called AiroPeek. This tool allows for the sniffing of wireless 802.11b networks. The latest version of AiroPeek supports 802.11b/g/a wireless technologies. The product is offered in two versions: SE and NX. The difference in these versions deals with supporting expert analysis modes and added advanced features for troubleshooting. This tool is commonly used for troubleshooting wireless networks. It allows the user to see what is going on from a frame level on the wireless network. Network sniffers have been around for some time now, and their usage has become an invaluable aid in troubleshooting complex problems. Another, more dangerous use of this tool is to step up a denial-of-service or man-in-the-middle attack. This sniffer program can provide several useful functions and information, although in the hands of a hacker, this can be a real risk.

AiroPeek provides numerous features.. It can support captures from multiple cards. This means one can set up two captures running on the same box and watch traffic in real-time go through different portions of a network. This is because the new version of AiroPeek supports the use of probes that it can connect to and use as remote sniffers.

Another interesting feature of AiroPeek is the ability to create trusted access points and clients. This makes the job of locating rogue access points a little easier. Once all trusted access points and clients are programmed into the sniffer, any non-trusted items will be flagged. AiroPeek does a great job of providing a large amount of decodes. Some of these involve complex Oracle and SQL functions to help troubleshoot database access problems. AiroPeek can also capture a large amount of VoIP, providing a great tool for identifying wireless VoIP issues.

Installing AiroPeek is straightforward: just download and execute setup.exe. This will prompt a wizard; go through the wizard, mostly selecting the defaults, and wait for the program to install. Once AiroPeek is installed on a computer, a second step is needed to start using the software. This second step involves installing an AiroPeek driver for the wireless card being used. In Windows, there are no open source programs to turn wireless cards into promiscuous mode. Because of this, AiroPeek had to create drivers for a number of common wireless cards. Most of these drivers ship with the software; however, for the most up-to-date list, check the AiroPeek Web site (www.wildpackets.com). Having located the AiroPeek driver for the particular wireless adapter in use, one can start to use the program.

Once the program is correctly installed along with the driver, it should look like Figure 16.5. Note the different functions available across the bottom. Each tab opens a different view on the capture, allowing for the use of different functions.

## 16.2.2 Sniffer Pro

Sniffer Pro is another type of wireless sniffer. It has the ability to capture wireless traffic in standard formats such as CAP. It also comes from one of the best-known sniffer companies, Network General. The latest version of

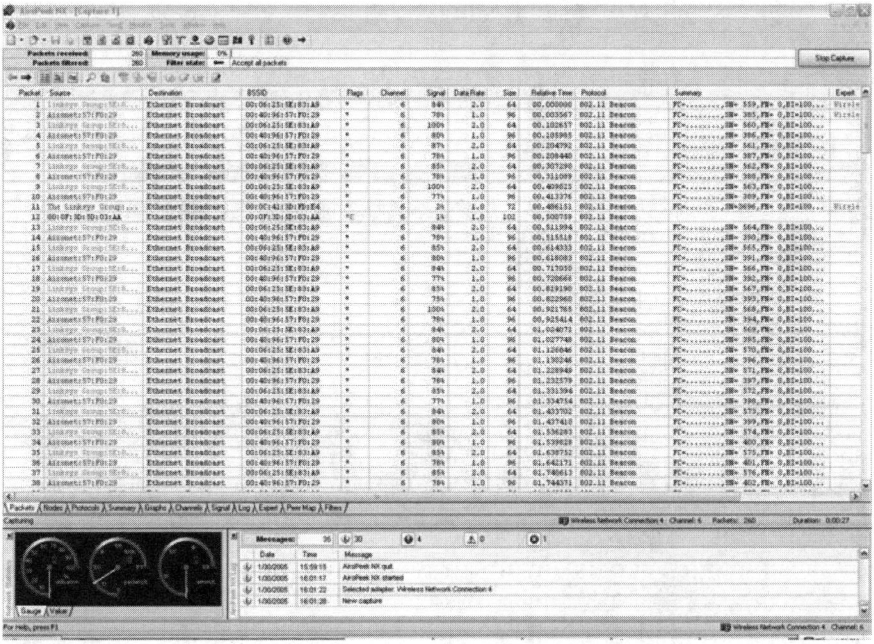

**Figure 16.5   AiroPeek screen shot.**

Sniffer Pro is capable of decoding 802.1b, 802.11g, 802.11a, and Bluetooth communications. This product does have some amazing capabilities, including replay, WEP key integration, and a distributed add-on feature. Looking at Figure 16.6, one can see what the sniffer program looks like.

### 16.2.3  Mognet

Mognet is a purely Java-written wireless sniffer. It can run on any device that supports Java and has a wireless card capable of monitor mode. It requires the Java Development Kit to compile one of the libraries that comes with it. Once one has the software, it is an easy install and can be quite handy as an easy-to-use wireless sniffer. Being Java makes Mognet a wireless sniffer that can operate on a large number of Java-capable devices. It includes real-time capturing and the ability to save files as ASCII or libpcap.

## 16.3  Hybrid Tools

When we defined wireless tools as scanners, sniffers, and crackers, we still needed some way to classify the tools that performed multiple

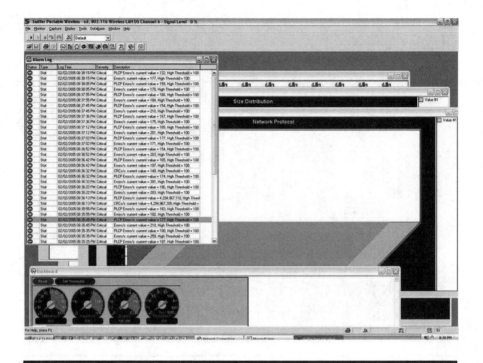

**Figure 16.6    Brutus screen shot.**

functions. This is why there is a hybrid section. Tools that can easily present a GUI (graphical user interface) with relevant wireless detection information are called scanners. If they have in-depth packet analysis features, they are sniffers. If they can display an easily readable GUI and have in-depth packet analysis under the hood, they are hybrid tools. Some of the most valuable tools in any wireless professional's toolkit are discussed in this section.

## *16.3.1 Kismet*

The Kismet tool was written by Mike Kershaw. This tool is a layer 2 wireless detector, sniffer, and intrusion detection system. It works on Linux and can detect, scan, and report on 802.11/a/b/g wireless networks. Just recently, in its newer versions, support was added for 802.11g and 802.11a. Kismet is able to find most of the items outlined in the wireless scanning section. It can detect the presence of a wireless network, if it is encrypted, what channel it is operating on, what MAC address it is using, and much more. A list of features from the Kismet read me file is given below. One thing that Kismet has that most other scanners do not is the ability to quickly point out if an access point is still running its default SSID; it also has the ability to detect Cisco access points via CDP. The Kismet tool is available form http://www.kismetwireless.net. To use it requires a Linux kernel as well as a supported wireless interface card. When the first version of Kismet came out, one of its biggest problems was trying to get the correct wireless card to work with it. Now in its most recent 4.0 release, it supports more than 20 different wireless cards. This has made Kismet easier to use, thus increasing it user base and making it a more popular tool.

Listed below are some of the features of Kismet from the Kismet read me file:

- Ethereal/Tcpdump compatible data logging
- AirSnort compatible weak-IV packet logging
- Network IP range detection
- Built-in channel hopping and multi-card split channel hopping
- Hidden network SSID decloaking
- Graphical mapping of networks
- Client/server architecture allows multiple clients to view a single Kismet server simultaneously
- Manufacturer and model identification of access points and clients
- Detection of known default access point configurations
- Runtime decoding of WEP packets for known networks
- Named pipe output for integration with other tools, such as a layer 3 IDS like Snort
- Multiplexing of multiple simultaneous capture sources on a single Kismet instance
- Distributed remote drone sniffing
- XML output
- More than 20 supported card types

Some of the more advanced features of Kismet include the ability to report on multiple packet sources. This means one can set up sensors to collect data from remote points and feed it back to a central Kismet console. It can feed this information into a number of other programs, such as AirSnort, Snort, Ethereal, and many others. It has the ability to export data as XML, HTML, and CAP files.

Looking at Kismet in action, one can see the full ability of this software. The Kismet program itself is a command-line tool. A GUI is available called ncurses. This front end displays the network information about a wireless network. Figure 16.7 shows what the Kismet program looks like and how it operates. It has a variety of useful messages. If it knows a network it found is a default SSID, it will set an F flag that identifies the access point as being a factory default. It has the ability to discover and

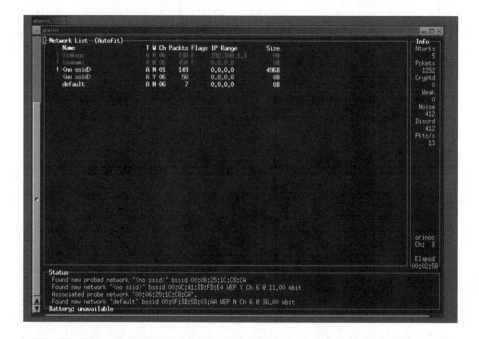

**Figure 16.7   Kismet screen shot.**

explain how it discovered network layer information about an access point. It uses the following flags to inform the user of how it determines this IP information:

- ARP for Address Resolution Protocol
- TCP for Transmission Control Protocol
- UDP for User Datagram Protocol
- DHCP for Dynamic Host Configuration Protocol
- CDP for Cisco Discovery Protocol

Kismet has some other GUIs available that were written by others. The first, called WireKismet, was primarily written for small screens, such as PDA devices. It has the ability to choose other Kismet servers to view and it supports a number of the features from the GUI. One of the most well-known front ends outside of the ncurse is called Gkismet. This front end is available at http://gkismet.sourceforge.net.

Kismet also has the ability to detect attacks being attempted on the network. Currently, it has a limited number of signatures, although they grow with each release. One of the more well-known signatures is the detection of someone using the Network Stumbler program. This program is one of the few wireless sniffers that can be detected. This is due to an active probe it sends out looking for additional information about a network. This is easily identified with an N flag. Some of the other signatures include:

- *Netstumbler fingerprinting.* Network Stumbler sends a unique packet in an attempt to identify the SSID of a network with SSID broadcast suppression enabled.
- *Deauthenticate/disassociate flooding attacks.* In such an attack, an attacker sends out disassociate or deauthenticate message that would knock legitimate users off the network.
- *Lucent tool testing.* This will repost if someone is using the Lucent, Orinoco, Proxim, or Agere site surveying tools on the network. This would be an initial step for a wireless hacker.
- *Wellenreiter.* If an attacker is using Wellenreiter and trying to find the SSID of a network with SSID broadcast suppression enabled, it will send out a unique probe packet that can be detected as the Wellenreiter software.
- *Access point channel change.* Kismet can detect if an access point has changed to a different channel. This can be used to know if an attacker has compromised the access point.
- *Broadcast deauthenticate/disassociate flooding attacks.* This is the same as a deauthenticate/disassociate flooding attacks, although with a broadcast, all clients will disassociate from the network.
- *AirJack tool detection.* When someone first uses the AirJack tool, it will set itself to the default SSID of AirJack. Kismet will report this if it is seen.
- *Excusive probing.* If a device probes for a network and does not connect after being accepted, this alert will sound. It will only alert after it has seen this happen more than once. This can help against tools that perform scanning.
- *Disassociating traffic.* If Kismet sees a device talk in less than ten seconds after it has disassociated, it is considered the victim of a disassociation attack or the attacker.
- *Sending zero-length probe response.* If a probe response is seen with no SSID, this alert will sound. Some manufactured radio cards have fatal errors when seeing a zero-length SSID.

As one can see, Kismet is a very versatile tool that can be used for both good and bad. Many commercial variants that only perform a limited set of the features of Kismet range upward of $10,000. Having a free tool that incorporates the entire feature set of Kismet is worth investigating.

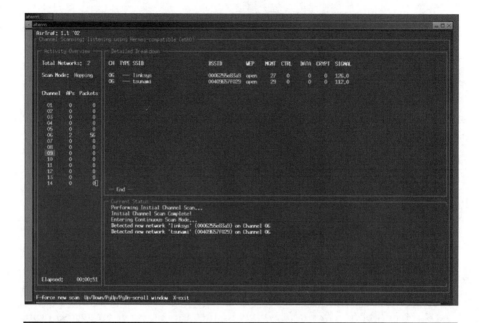

**Figure 16.8   AirTraf screen shot.**

## 16.3.2  AirTraf

AirTraf is a wireless sniffer that also incorporates wireless scanning functions. The AirTraf wireless sniffer is a free, open source program. The creators of AirTraf have formed a company called Elixar and have created a paid version of AirTraf for enterprise customers. The AirTraf program has packet capture/decode abilities just like other sniffers. It allows for the capture and analysis of management, control, and data frame types. It can also perform bandwidth calculation and signal strength identification. It allows for simple identification of the SSID, channel, MAC address, and bandwidth load for each access point it can detect. It can also track management traffic in a given area to determine whether or not the activity is hostile; this feature is new and still very buggy. The goal of the AirTraf program is to have a distributed wireless sniffer able to locate rogue devices, attacks, and general wireless load. The AirTraf program can run on any x86 Linux machine running a recent kernel release. Looking at Figure 16.8, one can see AirTraf in action. It has found some networks operating and it has listed if they are using WEP, what their SSID is, and each of their signal levels. This program can also perform a number of other functions such as sorting and filtering.

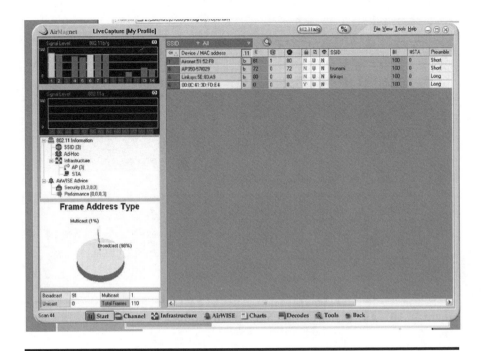

**Figure 16.9** AirMagnet screen shot.

## 16.3.3 AirMagnet

The AirMagnet tool can be used for a number of wireless troubleshooting tasks. Like all good network tools in the wrong hands, it can be rather evil. The AirMagnet tool operates on a Windows laptop or Compaq Ipaq. The software is expensive, making its use for the low-funded hacker less common. The software itself can sniff the airwaves, find noise, perform site surveys, locate rogue access points, and provide a high-level problem detection tool. The tool's main screen looks like the one in Figure 16.9; this main screen shows all the access points and clients that the device can currently detect. For further information on any particular device, a click is all that is needed.

Some of the tools that AirMagnet encompasses are shown in Figure 16.10. As one can see, there are a number of testing methods available within this software. Some of the more advanced tools are seen in Figure 16.9. These include multipath detection software and signal coverage check. In addition to these two screens, one of the more security-related abilities of this device is to create a white list of good access points. Once this is done, a quick manual scan can easily identify any rogue access points.

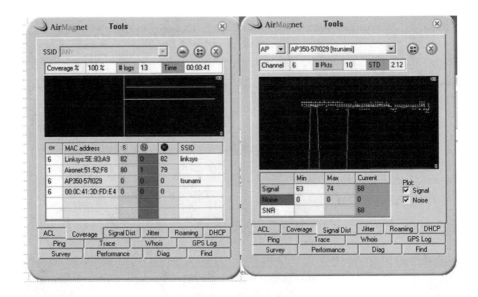

**Figure 16.10  AirMagnet tool section screen shot.**

# 16.4  Denial-of-Service Tools

This section discusses some of the tools that perform attacks on availability of wireless networks. Many threats can hinder the ability of a wireless network to work. The most common of these threats is the fear of signal jamming. Jamming is the biggest threat facing wireless in general; when looking at WLAN in particular, one sees that another form of preventing service is possible. This is a DoS (denial-of-service) attack. When this happens, its effects are similar to a jamming attack; however, the frequency is not jammed, just the management logic inside the wireless device. These threats stem from the poor setup of secure management traffic on IEEE wireless networks. It is very easy to create a DoS attack on a wireless network. In this section, we look at some automated tools used to create DoS attacks. As pointed out in the "Breaking Wireless Security" chapter (Chapter 13), a wireless sniffer with replay option can create the same type of attack.

## 16.4.1  WLAN-Jack

This tool is an automated tool to send disassociation messages onto a wireless network. Its objective is to knock a wireless client off his or her network and force them to re-authenticate. If this re-authentication happens, the hacker gets the chance to capture the authentication or perform a man-in-the-middle attack. These are the most common reasons why a wireless hacker would perform a DoS attack.

### 16.4.2 FATA-Jack

This tool is similar to WLAN-Jack, although it performs another type of DoS attack. It was found that a malformed authentication packet with the algorithm set at type 2 along with a sequence number and status code both of 0xffff could create a DoS result in access points. This combination would make the access point re-authentic a wireless client, thus creating a DoS attack and allow the attacker to see the authentication process.

## 16.5  Cracking Tools

The wireless cracking tools discussed in this section are used to break network encryption types on wireless networks. They work by taking advantage of the encryption type or the method by which the encryption is applied. Most ciphers have strong cryptographic functions; however, the mechanisms used to implement the cipher often have flaws that are identified and then exploited. This section identifies some of the widely used cracking tools for wireless networks.

### 16.5.1 WEPCrack

The WEPCrack tool was the first tool to crack WEP (Wired Equivalent Privacy). It was created before the more famous AirSnort tool. Currently, the WEPCrack tool is maintained by Anton T. Rager and is available at http://wepcrack.sourceforge.net. WEPCrack is a couple of Perl scripts that validate the workings of Fluhrer, Matin, and Shamir's (FMS) theoretical attack. It took their paper and created an automated tool to perform and prove their research.

These Perl scripts are as follows:

- WEPCrack.pl
- WeakIVGen.pl
- Prism-getIV.pl
- Prismdecod.pl

The tool works by collecting weak IVs and sending them to the IVfile.log. Once they are in the log file, the WEPCrack.pl can run the FMS

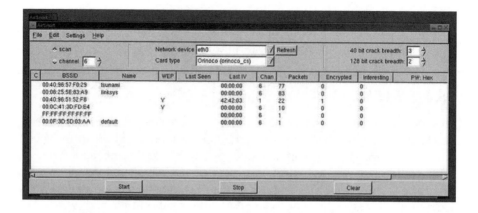

**Figure 16.11    AirSnort screen shot.**

attack against them. The WEPCrack tool uses pcap files and Perl. This allows the program to be very versatile; it can be loaded on a PDA or another device that supports Perl. This tool does not have a front end or a pretty GUI. This helps with battery life and processing power, although it affects the ability to run the program by novice users. This is one of the main reasons that the next tool (AirSnort) has become the more popular choice for WEP cracking.

## 16.5.2 AirSnort

The AirSnort tool was created by the Shmoo group. It is a WEP cracking tool with a nice GUI or front end. This has helped it become the tool of choice for many wireless auditors, hackers, and curious people. This tool has the ability to take pcap files similar to WEP crack. This gives the user of the tool the ability to use another tool (or device altogether) to collect the data. This is very useful when performing audits; no one wants to lug around a heavy high-powered laptop. Using a PDA, an auditor can collect the needed data and then use a high-powered desktop to perform the cracking. The AirSnort tool also uses the method discovered by FMS.

In Figure 16.11, one can see AirSnort in action, cracking a number of wireless networks. The user interface is rather easy to operate. One selects the adapter and its corresponding chipset and then hits start. One can also play with the *breadth* command to make cracking attempts a little faster. All the networks found are listed in the window showing how many packets were seen and, of those, how many have WEP enabled. Once a network is found with WEP enabled, the AirSnort program starts to look for interesting IVs. Every time one of these IVs is found, the program performs the FMS attack on it in hopes of recovering a key byte.

Every time this happens, the IV counter on the program gets incremented and the WEP key becomes a little closer to being cracked.

### 16.5.3 BSD-Airtools

BSD-Airtools is a number of programs similar to WEPCrack. As the name implies, BSD-Airtools runs on the Berkeley Software Distribution (BSD) platform. The tools are dwepdump, dwepcrack, and dwepkeygen. The dwepdump tool is used to find interesting IVs and send them to a location where dwepcrack performs a number of attacks on them. BSD-Airtools uses the FSM attack. It also has created a faster method based on patterns that interesting IVs provide. For a more detailed look, visit the BSD-Airtools Web page at http://www.dachb0den.com.

### 16.5.4 ASLEAP

The ASLEAP tool was created by Joshua Wright to attack Cisco LEAP. The tool works by taking advantage of a weakness in the Cisco LEAP wireless security method. This tool is currently in version 1.4, which was released December 17, 2004. This new version not only has the ability to passively collect and extract LEAP usernames and passwords, but has now added support for the Point-to-Point Tunneling Protocol (PPTP). So far, this tool can only find passwords that are part of a already-existing word list. This makes strong passwords that take into account complexity requirements safe from the dangers of an attacker running ASLEAP on a Cisco LEAP-enabled wireless network.

## 16.6 Access Point Attacking Tools

This section details the tools that one can use to attack the access point itself. These tools are used to exploit many widely used Internet protocols for remote management and monitoring. Each of these tools is not directly related to wireless, although they will still defeat the security of an access point. Most of these tools take advantage of problems that exist in widely used protocols such as SNMP, HTTP, and Telnet. This section provides the tools used to perform the attacks outlined in Section 13.3.

### 16.6.1 Brutus

Brutus is a free remote password cracking utility. This software was released in 1998. Since then, it has been downloaded from the hoobie

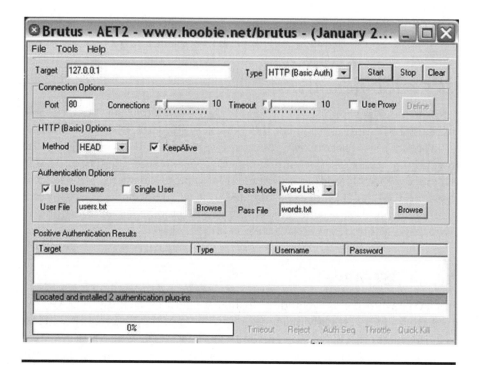

**Figure 16.12   Sniffer Pro screen shot.**

Web site more than 7000 times. The current version of Brutus is called AET2; it is capable of working on many versions of Windows, including Windows 9x, NT, 2000, and XP. It can attempt wordlist, dictionary, and brute force attacks on many Internet protocols. Some of the protocols supported in Brutus are HTTP, POP3, SMTP, FTP, SMB, Telnet, NNTP, and IMAP. Other applications and protocols can be used with added Brutus Application Definition files. These files allow users to create special-case applications they want to test. These files are easy to import and export, making Brutus a common attack tool. This tool is widely used and some anti-virus software packages have a signature for this tool, marking it as a hacker tool.

Figure 16.12 illustrates Brutus performing an attack. As one can see from the upper right-hand corner, a number of protocols can be set for Brutus to crack. The two that are most interesting with respect to wireless access points are Telnet and HTTP. Once a protocol is chosen, the next step is to select a wordlist or a username list. This software has the ability to try a statically defined username or a list of multiple usernames. Once this has been selected, a word list must be inserted, or one can use the rather small wordlist that comes with the program itself. One of the largest wordlists to date is located on the AE&E Corporation Web site. It has a file size of 14 megabytes. As a flat text file with 14 megabytes, one can be sure it accounts for most all words, names, places, and other groupings of letters often used for passwords. Once both the username and password sections have been completed, all that is needed is to pick a target and run the utility. One note here is that some newer access point versions have realized the vulnerability this program presents and have modified their software to protect against attacks like this.

## 16.6.2 SolarWinds

SolarWinds (www.solarwinds.net) is a set of programs used for network management. The full version has support for 45 tools that can help identify security and performance issues within large networks. These independent tools make up the SolarWinds bundle. This software bundle is available in four groups. Each group allows the customer to purchase more and more tools, from the lowest version with seven tools to the highest version with all 45 tools. These tools are only sold in the bundle; and while they may operate on their own, they are not sold separately. Of the available 45 tools, the discussion here focuses only on a few security-related tools that can be used to attack access points. Each of the security-related tools that SolarWinds ships with is listed below, along with a description of what they do.

### 16.6.2.1 Port Scanner Tool

The Port Scanner tool allows for testing open TCP ports on devices. It can be set up to run across a single IP address or a range of IP addresses. It has the ability to locate certain port ranges on machines across large networks. This tool allows for easy reporting and fast operation. Figure 16.13 shows this tool in action. From the figure one can see that it defaults to some common port numbers to allow for a quicker scan. Once the scan takes place, not only does it scan for open ports numbers, but it also tries to find device names through DNS.

### 16.6.2.2 SNMP Brute Force Attack Tool

The Brute Force Attack tool will attack an IP address with multiple SNMP queries to try to determine the SNMP read-only and read-write community strings. It accomplishes this by trying every possible community string combination. Figure 16.14 shows this program in operation. As one can see, it can attack an IP address and try all possible SNMP combinations to find the SNMP string.

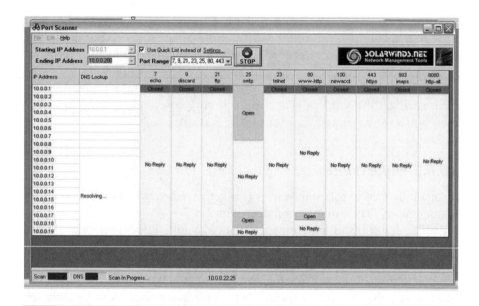

Figure 16.13   SolarWinds port scanner screen shot.

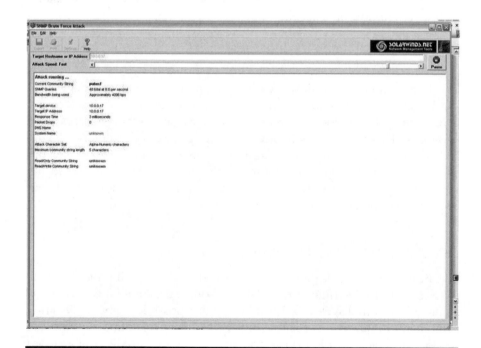

Figure 16.14   SolarWinds SNMP brute force attack scanner screen shot.

**Figure 16.15    SolarWinds SNMP dictionary attack scanner screen shot.**

### 16.6.2.3  SNMP Dictionary Attack Tool

This tool performs the same attacking function as the Brute Force SNMP tool. The main difference between the two is that the Brute Force Attack tool uses every possible combination, whereas the Dictionary Attack tool uses a preset list of words, phrases, or numbers. Looking at Figure 16.15, one sees that this tool takes an access database of IP addresses and applies another wordlist of SNMP strings to try on a device. This tool works remarkably fast in finding the correct SNMP strings on a network device.

### 16.6.2.4  Router Password Decryption Tool

This tool is not widely used on wireless; however, if the wireless access points are Cisco, this tool can perform some useful features. This tool is a decrypting tool for any Type 7 Cisco passwords. These passwords are on the devices configuration as an encrypted hash of the administrator's password. If one can look at the configuration of an access point with SNMP read string, one can copy this value and use this tool to decrypt the password. Figure 16.16 shows this program in action. The password that was originally entered into the Cisco configuration is the phrase "this is crackable." As one can see from Figure 16.16, this program makes little work of decrypting the password, even with a long phrase entered.

## 16.6.3  Cain and Able

This tool is like the "Swiss Army knife" of authentication crackers. It can crack MD5 Cisco hashes, RADIUS authentication, RSA Secure IDs, and many other types of authentication. It allows for easy recovery of passwords by using the built-in sniffer to capture passwords that can be fed to the cracking engine. This engine can crack encrypted passwords using a dictionary, brute force, or cryptanalysis attack. It can also decode scrambled passwords, revealing password boxes and uncovering cached

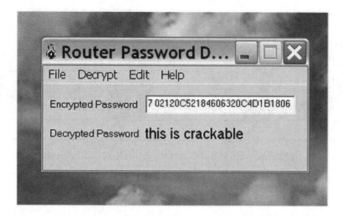

Figure 16.16  SolarWinds router password decryption tool scanner screen shot.

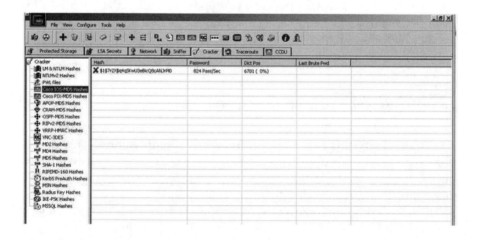

Figure 16.17  Cain and Able screen shot.

passwords. It also has the ability to analyze routing protocols. It runs on Windows and is available from the following Web site: http://www.oxid.it.

Figure 16.17 illustrates the Cain and Able tool cracking a password. This screen shot only shows the authentication types on the left that can be fed to this program. This program has the ability to pull this information right out of the wire with its built-in sniffing capability. This makes it a real threat to an advanced user on a wireless or wired network. One other interesting note about this program is that when it first starts up, it will find all Internet Explorer cached credentials used to access a Web page. This might be something that you might not want to open in front of too many people until you have looked at it yourself. Another interesting use of this program is at public terminals; because it shows all cached usernames and passwords, it will display some juicy information if someone mistakenly cached his Web credentials on this terminal.

## 16.7 Other Wireless Security Tools

Some tools that do not fit into the four main groups listed above can be used on wireless networks. To gain a full understanding of what tools are out there and what types of security methods they can defeat, one needs to look at the tools identified in this section. These tools detail how to circumvent MAC-based filters and SSL gateways. This section looks at each of these tools on their own, explaining where they came from, what they are for, and how they work.

### 16.7.1 EtherChange

EtherChange is one of many programs on the market to change the MAC address of machines. In UNIX and Windows, one can change the MAC address through the operating system without the help of a tool. These tools just make it easier for a novice to perform the same actions. To use the tool, all that is needed is the executable that can be downloaded from the Internet. Once this executable runs, it looks like Figure 16.18. This

```
C:\Documents and Settings\aearle\Desktop\etherchange.exe                 _ □ ×

EtherChange 1.0 - (c) 2003, Arne Vidstrom
             - http://ntsecurity.nu/toolbox/etherchange/
0. Exit
1. Intel(R) PRO/100 VE Network Connection
2. Cisco Systems 350 Series PCMCIA Wireless LAN Adapter
3. 1394 Net Adapter
4. Sierra Wireless AirCard 555 Adapter
5. AirMagnet Aironet 802.11a/b/g Wireless Adapter

Pick a network adapter: 2

0. Exit
1. Specify a new ethernet address
2. Go back to the built-in ethernet address of the network adapter

Pick an action:
```

**Figure 16.18   EtherChange screen shot.**

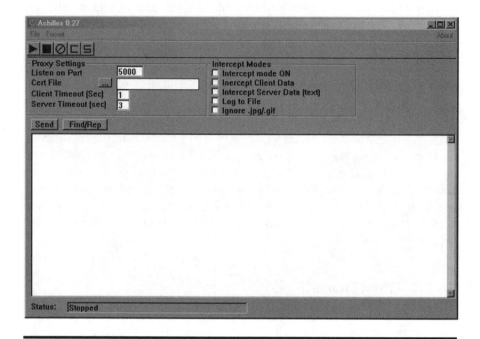

**Figure 16.19   Achilles screen shot.**

shows all available network cards and allows the user to change the MAC address of any of these adapters. Once the MAC address has been changed, the network card must be reset or the machine must be rebooted. Once the card has reinitialized, it has the selected MAC address and not the one that originally came with the card. This can be used to circumvent MAC-based filtering.

## 16.7.2  Achilles

The Achilles tool is used testing the security of Web applications. It can also function as a proxy server, which acts as a man-in-the-middle during an HTTP session. This can also be applied to SSL, allowing a man-in-the-middle attack to occur on SSL wireless gateways. This software runs on the Windows platform and supports Windows NT, Windows 2000, and Windows 98. This can intercept an HTTP and SSL session by acting as an invisible proxy server. This gives the Achilles user the ability to look, log, and change data in transit from a client to a Web site. On a wireless gateway, this would allow an attacker to set this up and perform a man-in-the-middle attack. Once this happens, the user would not even know that he is not connecting to the gateway. His session is really going to the gateway, although the SSL encrypted session is decrypted in the Achilles program and another SSL session is established between Achilles and the wireless gateway.

Figure 16.19 shows the Achilles program. As one can see, it has the ability to listen on multiple ports and can include a cert to provide SSL capability.

## 16.8 Chapter 16 Review Questions

1. What type of tool is Network Stumbler?
   a. Wireless sniffer tool
   b. Wireless scanner tool
   c. Wireless cracker tool
   d. Wireless DoS tool
   e. Wireless hybrid tool

2. Wireless scanners all perform sniffing functions in the background.
   a. True
   b. False

3. What type of tool is AiroPeek?
   a. Wireless sniffer tool
   b. Wireless scanner tool
   c. Wireless cracker tool
   d. Wireless DoS tool
   e. Wireless hybrid tool

4. Which of the following tools are not free of charge?
   a. AirSnort
   b. Network Stumbler
   c. AiroPeek
   d. AirMagnet

5. What type of tool is AirSnort?
   a. Wireless sniffer tool
   b. Wireless scanner tool
   c. Wireless cracker tool
   d. Wireless DoS tool
   e. Wireless hybrid tool

6. What type of tool is MiniStumbler?
   a. Wireless sniffer tool
   b. Wireless scanner tool
   c. Wireless cracker tool
   d. Wireless DoS tool
   e. Wireless hybrid tool

7. Wireless scanners are used for advanced troubleshooting.
   a. True
   b. False

8. What type of tool is Kismet?
   a. Wireless sniffer tool
   b. Wireless scanner tool
   c. Wireless cracker tool
   d. Wireless DoS tool
   e. Wireless hybrid tool

9. What type of tool is AirMagnet?
   a. Wireless sniffer tool
   b. Wireless scanner tool
   c. Wireless cracker tool
   d. Wireless DoS tool
   e. Wireless hybrid tool

10. Which of the following tools are free of charge?
    a. AirSnort
    b. WEPCrack
    c. AiroPeek
    d. AirMagnet

11. Which of the following tools do not crack WEP keys?
    a. AirSnort
    b. WEPCrack
    c. BSD-Airtools
    d. Kismet

12. What type of tool is WLAN-Jack?
    a. Wireless sniffer tool
    b. Wireless scanner tool
    c. Wireless cracker tool
    d. Wireless DoS tool
    e. Wireless hybrid tool

13. What feature does AiroPeek and AirMagnet share?
    a. Remote sniffing
    b. WEP cracking
    c. Rogue access point identification
    d. Packet replay

14. Kismet can be used for a WIDS.
    a. True
    b. False

15. Network Stumbler works on what platforms?
    a. Linux
    b. UNIX
    c. Windows 2000
    d. Windows XP

# Appendix A

# Review Question Answers

## Chapter 1 Review Answers

1. What happens to an 802.11b wireless signal when an evergreen tree is located between the transmitter and receiver?
   e. It is absorbed.

2. What are the two correct terms used to measure antenna gain?
   c. dBD and dBi

3. The designator dBi is a decibel compared to _____
   c. Isotropic radiator

4. What does RF stand for?
   Radio frequency

5. Which type of modulation does 802.11b use?
   d. CCK

6. How can one send more data across the air?
   b. Use a more complex modulation.

7. What was the Wi-fi Alliance formerly known as?
   b. WECA

8. What seal certifies interoperability in a manufacturer's wireless device?
   d. Wi-Fi Certified™

9. Which wave will travel the greatest distance?
   a. FM radio

10. What two items should be maintained near the edges of a wireless cell when performing a site survey?
    a. High signal-to-noise ratio
    d. High signal strength

11. What would the FCC and ETSI regulate on a wireless network? (Select more than one)
    a. Power outputs
    c. Channel number and frequency

12. What bandwidth term is this phrase stating? On any given day, my wireless network has a pretty low bandwidth of _____.
    a. 11 Mbps

13. Which of the following shows the correct use of a wireless network?
    b. Mobile access from laptop or PDA

14. When a wireless signal changes or bends around an object, sometimes creating a shadow area, it is known as _____.
    c. Diffraction

15. When doing wireless power calculations, what two terms are often converted into each other?
    d. dBd to dBi

16. What standards body creates wireless standards?
    b. IEEE

17. What key size is required in WEP for the Wi-Fi Alliance to certify a product?
    a. 40 Bit

18. What does Wi-Fi stand for?
    c. Wireless fidelity

# Chapter 2 Review Answers

1. Which of the following processes will not lower the risk of social engineering at a help desk?
   d. Having the caller verify the identity of the help desk operator

2. What would a hacker whose motive was money most prefer to attack?
   b. Bank

3. Which of the following terms best describe a Wi-Fi hacker?
   d. War driver

4. What type of malicious code infects devices and does not have the ability to replicate or spread outside the infected system on its own?
   b. Virus

5. List the three main goals of information security.
   a. Integrity
   c. Availability
   d. Confidentiality

6. What two pieces of information are required to hack a cell phone?
   b. MIN and ESN

7. What piece of information is unique on every wireless card in the world?
   c. MAC address

8. Which of the following terms are used to describe the process of discovering wireless networks?
   d. All of the above

9. A self-replicating and often self-sending piece of malicious code, which is often e-mailed, is called _____.
   a. A worm

10. What technique would an attacker do to force a wireless end device to disconnect from the network?
    d. RF jamming

11. What would a wireless hacker in his early teens most likely be doing?
    c. War walking

12. What does the term "spam" mean?
    b. Unwanted, unsolicited bulk electronic messages

# Chapter 3 Review Answers

1. What piece of legislation defines communication as "the transfers of signs, signals, writing, images, sound, data, or intelligence of any nature transmitted in whole or in part by wire, radio, electromagnetic, photo-electronic, or photo-optical systems that affects interstate or foreign commerce"?
    b. Electronic Communications Privacy Act

2. What act detailed the use, disclosure, interception, and privacy of electronic commutations?
    b. Electronic Communications Privacy Act

3. How much loss must a company have before it can get help from the FBI?
    d. It is based on location and the average loss for that location.

4. What did the Patriot Act amend relating to computer crime? Choose the best answer.
    b. The Electronic Communications Privacy Act

5. When was the Electronic Communications Privacy Act created?
    d. 1986

6. When was the Computer Fraud and Abuse Act created?
    d. 1996

7. What act was created to help protect U.S. Government computers?
    c. Computer Fraud and Abuse Act

8. What act was created to protect U.S. networks?
    c. Computer Fraud and Abuse Act

9. Under the Patriot Act, who was required to log and track hacking attempts?

   c. The ISP

10. What regulation, act, or law was put into place due to the lack of clear terms regarding what constitutes a computer crime?

    c. The Computer Fraud and Abuse Act

11. The FCC has a law that makes running a sniffer and receiving other people's network traffic a crime.

    a. True

## Chapter 4 Review Answers

1. An 802.11a radio uses what technique to transmit its signal?

   c. OFDM

2. The 802.11b standard does not use DSSS.

   b. False

3. What standard listed below supports Frequency Hopping Spread Spectrum?

   d. 802.11

4. Within North America, what number of channels are considered nonoverlapping?

   b. 3

5. The 802.11g standard supports the use of orthogonal frequency division multiplexing in which radio frequency band?

   a. 2.4 MHz

6. Dwell time is a function of which of the following physical layer techniques?

   b. FHSS

7. DSSS is a type of what?

   b. Spread spectrum

8. How many channels of DSSS are used in the United States?
   d. 11

9. How many channels are used in an ESTI-governed location?
   b. 13

10. The 2.4 frequency that 802.11b operates in is called _____.
    d. ISM

11. The 802.11a standard is associated with which two terms?
    b. UNII
    d. 5.15–5.30 GHz

12. How much of the 2.4-GHz ISM band is used when a single access point 802.11b wireless network is enabled?
    d. 20 MHz

13. What is the total amount of wireless bandwidth inside the 2.4-GHz ISM band?
    d. 83.5 MHz

14. What is the amount of time a channel takes to switch from one channel to another channel on an FH system called?
    a. Hop time

15. You are designing an RF network and have channel 6 used. What other channel would you select to use next to channel 6?
    c. 1

# Chapter 5 Review Answers

1. What are the three main wireless frame types?
   a. Control frame
   b Management frames
   c. Data frames

2. When a wireless network is using an access point, which device sends the beacons?
   a. Access point

3. Which layer 2 protocol does 802.11b use?
   d. CSMA/CA

4. What is the difference between a Disassociation and a De-authentication frame?
   c. Reason code

5. CSMA/CA is a _____ process.
   d. Four-step

6. To perform management functions and allow for time synchronization, a preset interval where nothing communicates is required. What term does this describe?
   d. Interframe spacing

7. Which step takes place first: authentication or association?
   a. Authentication

8. CTS/RTS is part of _____?
   a. CSMA/CA

9. Which layer 2 protocol does Ethernet use?
   a. CSMA/CD

10. What is the SSID used for?
    a. Identification

11. What distributed coordination function (DCF) is used to _____.
    c. Poll the media for network availability

# Chapter 6 Review Answers

1. The 802.11 standard provides what maximum data rate?
   b. 2 Mbps

2. With Ad Hoc mode, an AP is required.
   b. False

3. Most wireless networks operate in what mode?
   b. Infrastructure

4. The 802.11b standard provides what maximum data rate?
   a. 11 Mbps

5. What part listed below radiates a wireless signal?
   b. Antenna

6. Which of the following terms do 802.11/b/g/a have in common?
   b. CSMA/CA

7. In the United States, what is the maximum EIRP limit on 802.11a UNII 1 system?
   b. 100 mW

8. The 802.11a standard provides what maximum data rate?
   d. 54 mbps

9. What UNII band is for outdoor use?
   c. UNII 3

10. The 802.11g standard provides what maximum data rate?
    d. 54 Mbps

11. Which of the following PAN technologies does 802.11 address in its standard documentation?
    b. IR

12. What happens on an 802.11g network when an 802.11b client is injected?
    c. The access point and all other access points that can hear the client will downshift to 802.11b.

13. What standard would one need to install Wi-Fi phones in a network? Select the best answer.
    c. 802.11e

14. What is the newly formed draft standard that will replace 802.11g?
    c. 802.11n

15. Which one of these terms is not part of the wireless local area network service set?
    d. IESS

16. Roaming is defined in which of the following standards?
    d. None of the above

## Chapter 7 Review Answers

1. What type of cellular communications is considered G1?
   c. AMPS

2. What type of cellular communications is considered a global standard?
   b. GSM

3. What is the data rate of the CDMA2000?
   b. 144 kbps

4. The 802.16 standard has built-in mobility.
   b. False

5. GPS has how many satellites orbiting the earth?
   a. 24

6. What cellular standard was held back for many years as a military-only technology?
   d. CDMA

7. Which of the following terms is the one that finds other cellular users?
   a. Gateway mobile switching center (GSMC)

8. What is the documented as well as the real data rate of GPRS? Select two.
   a. 170 kbps
   d. 55 kbps

9. What is the data rate of the 1xEV-DO?
   d. 2 Mbps

10. Which cellular standard uses slices of time to allow for multiple connections?
    a. TDMA

11. The A3, A8, and A5 algorithms are part of what standard?
    b. GSM

12. All the research performed on the security of GPS has led the U.S. Government to say that the biggest risk to the GPS system is _____.
    b. Jamming

13. The 802.16e standard addresses mobility.
    a. True

14. What body performs interoperability testing for 802.16?
    c. WiMAX

15. What is the major difference between 802.16 and 802.20?
    c. Ability to receive a signal at high rates of velocity

16. What is the data rate of the TDMA?
    c. 64 kbps

## Chapter 8 Review Answers

1. When radiating an RF signal to create an area of coverage, what part of the WLAN is used?
   c. Antenna

2. What happens to the radiation pattern of an antenna when the gain is increased?
   a. The angle of radiation becomes larger.

3. Which antenna allows for coverage in a 180° area?
   b. Patch

4. What happens when a 50-foot antenna cable must be replaced with a 100-foot cable?
   b. The area of coverage will shrink.

5. Which antenna allows for coverage in a 360° area?
   c. Omni

6. EIRP is a representation of the power that an entire RF system has. When correctly measuring EIRP, what should be measured?
   c. Antenna output

7. What connector is used on a 2.4-GHz 1200 Series Cisco access point?
   c. RP-TNC

8. How does one calculate EIRP?
   c. Transmitter power + Antenna gain − Cable loss

9. To help with null areas and multipathing, what antenna architecture would you recommend?
   d. Diversity

10. Some access points have two antennas on them. Why? (Select two)
    b. To reduce the effects of multipathing
    d. Antenna diversity

11. There is a customer who wants to connect one office building to four warehouses in the area. Most of the offices are located less than a block away; the farthest one is half a mile away. Placing what antenna on the roof of the office will provide the best coverage?
    d. 5.2-dBi Omni

12. What is the designator for dBi? (Select two)
    b. Decibel
    d. Isotropic

13. The area around or immediately surrounding a wireless signal between two locations is known as the _____.
    c. Fresnel zone

14. What is the typical line of sight when using wireless between two buildings?
    d. 6 miles

15. You are trying to provide coverage down a long hallway. What is the best antenna solution for this?
    a. Yagi

16. Determine the EIRP for the following configuration: a 20-dBm radio using a 13.5 Yagi antenna with a 100-foot cable at a loss of 8 dBi per 50 feet.

   d. 29.5

17. You are connecting two buildings together with a wireless bridge. You are using a Cisco bridge with a Yagi antenna. In between the two buildings are some trees. You are unable to correctly set up a connection. What should you do?

   c. Raise the antenna above the treetops.

18. If an antenna has 6 dBd rating, what is that rating in dBi?

   d. 8.14

19. When an antenna gain increases, its beamwith _____.

   b. Decreases

20. Which antenna is most directional?

   d. Parabolic dish

21. What is the designator for dBd?

   a. Dipole

# Chapter 9 Review Answers

1. What is the second step in a wireless deployment process?

   d. Estimation

2. When performing a survey, what should the surveyor NOT do?

   d. Use local power

3. When performing a survey, what should a survey team always assume?

   b. Everything will break

4. When creating a business case, what piece of data is most important?

   c. Estimated cost

5. When creating an RFQ for a site survey, when should the warranty of the site survey end?
   c. Certification

6. What is the first step in a wireless deployment process?
   a. Business case

7. What is the final step in a wireless deployment process?
   c. Audit

8. What item is not normally included in a site survey kit?
   b. Telescoping pole

9. What is a spectrum analyzer used for?
   b. To locate interference in the 2.4-GHz range

10. Which item inside the physical information document is not correct?
    c. IP address of access point

11. What step of the deployment process is used to validate the site survey vendor's coverage and work?
    c. Certification

# Chapter 10 Review Answers

1. What is the correct DC voltage used to power a Cisco 1200 series access point when using POE?
   d. −48

2. The cable used to connect to a Cisco 1200 series access point is commonly referred to as a _____.
   a. Patch cable
   b. Coaxial
   c. Roll down
   d. Roll over

3. What administrative methods are supported in a Cisco access point?
   b. Console
   c. Telnet
   d. Web based

4. Which of the following is NOT a connection on the Cisco 1200 series access point?
   c. Fiber

5. The Cisco 1100 series access points support which antenna?
   c. 2.2 dBi Omni

6. What administrative methods are supported in a Linksys access point without firmware modifications?
   d. Web based

7. What command should be issued on a Cisco IOS access point after accessing it from a Telnet or console to reach the enable mode?
   c. Enable

8. What types of operating systems do new Cisco access points use?
   a. IOS
   b. VxWworks

9. Do Linksys access points support POE?
   b. No

10. Will Cisco sell you 350 series access points today?
    b. No

11. Cisco 1200 Series access points are capable of supporting 802.11b, 802.11g, and 802.11a all at the same time.
    a. True

12. What standard do all Cisco 350 series access points adhere to?
    b. 802.11b

13. What Cisco access point runs the VxWorks operating system?
    c. 350

# Chapter 11 Review Answers

1. What program is NOT in CE.NET but is in Pocket PC?
   c. Outlook

2. What was one of the first PIM devices?
   d. Zoomer

3. What is one of the biggest risks facing wireless devices?
   a. Theft

4. Healthcare has seen a great increase in productivity due to which wireless end device?
   b. Tablet

5. Windows' first version of CE was based on _____.
   c. Windows 95

6. What technology do most smart phones have that can be used to track them?
   c. GPS

7. When was the first computer virus written?
   c. 1983

8. What security feature has Microsoft put into its Smartphone OS and PocketPC OS?
   b. User action required to run an executable file

9. What makes securing a wireless end device more difficult?
   b. Multiple connection options (e.g., WLAN, PAN, cellular)

10. What does the acronym "LBS" stands for?
    c. Location-based services

11. Which operating system was created from the ground up as a PDA, phone, and PIM combo?
    c. Symbian

12. What is a major reason why handheld scanners have not gone 802.11g?
    c. Battery life

# Chapter 12 Review Answers

1. Which encryption mode was selected for 802.11i?
   c. AES-CCMP

2. Which EAP type is part of the EAP RFC?
   b. EAP-MD5

3. What EAP method did Microsoft, Cisco, and RSA create?
   a. PEAP

4. Does China have its own wireless security standard?
   a. Yes

5. In 802.1x, what is the RADIUS server often called?
   c. Authentication server

6. In 802.11i, what mode is one using if WEP is enabled?
   b. TSN

7. Which IEEE standards deal with security? (Select two)
   c. 802.11i
   d. 802.1x

8. During the time frame when WPA was released and 802.11i was still being finalized, what reservations would an IT manager have about using TKIP?
   b. Lack of interoperability

9. Which EAP type is used to create a tunnel in which other, older authentication methods can take place?
   d. EAP-TLS

10. Which of the following is not a VPN technology term?
    b. SSL

11. Which of the following provides layer 2 security by allowing for unique and changing encryption keys?
    c. TKIP

12. During an EAP negotiation, what happens after the identity response packet is sent from the client to the access point?
    a. The RADIUS server sends out a challenge.

13. What is a new federal encryption standard that is used in a certain mode on wireless networks?
    d. AES

14. What is used to ensure the integrity of packets traveling across the airwaves?
    c. MIC

15. WEP keys can also be thought of as _____.
    d. Shared secrets

16. Most wireless equipment is set to which better authentication method?
    a. Open key authentication

17. Which of the following can be used with the 802.1x standard for wireless authentication?
    b. RADIUS

18. A WEP key is _____.
    c. Weak

19. What is EAP-FAST replacing?
    b. LEAP

20. What IEEE group is responsible for creating WLAN security standards?
    d. 802.11i

21. Which of the following is not a correct EAP type?
    d. EAP-TTTLS

20. What security solution provides for unique changing keys that change to prevent an attacker from cracking the key?
    c. Dynamic WEP

23. What wireless security cipher is approved by the government for standard use?
    c. AES

24. What IEEE standard provides the authentication framework for all IEEE 802-based networks?
   b. 802.1x

25. What EAP type creates an encrypted tunnel and then performs another EAP exchange inside this tunnel?
   b. EAP-TLS

# Chapter 13 Review Answers

1. What is the process of scanning devices for open ports called?
   b. Enumeration

2. Using a corrupt EAPOL packet creates what type of attack?
   c. Denial-of-service attack

3. Why would an attacker use another wireless technology outside what the target is using to place a rogue access point?
   d. To prevent rogue detection software from finding it

4. What was the most significant attack on 802.1x?
   b. Man-in-the-middle attack

5. When an attacker sends a piece of information that is encrypted back into the network in hopes that the access point will decrypt it, what attack is performed?
   c. Double encryption attack

6. SNMP can be used to reset passwords.
   a. True

7. When using RADIUS, how many characters should the shared secret be?
   d. 24

8. 802.11i with preshared keys provides no current security risks.
   b. False

9. What attack on WEP are most WEP cracking tools based on?
   e. Stream cipher attack

10. Telnet provides a secure means of remote authentication.
    b. False

11. When using a wireless gateway, what protocol is most likely to become compromised?
    c. SSL

12. The process of using Google to locate relevant data about a target is performed at what step in the hacking process?
    a. Information gathering

13. Name two ways an SSID can be captured.
    a. Capturing probe frames
    b. Capturing beacon frames

14. To compromise a key, which of the following are needed? (Select two)
    a. Cleartext packet
    c. WEP encrypted packet

15. What would be the most likely place to find out about a company's IT equipment?
    c. Checking help forums for a company's name

# Chapter 14 Review Answers

1. What is the first step in developing a wireless security policy?
    b. Perform a risk assessment

2. What process results in the identification of flaws in the wireless network?
    c. Security audit

3. What types of companies should have a wireless security policy?
    d. All companies

4. What is the most important factor in creating a wireless security policy?
    b. Management buy-in

5. Which document does not go into detail on how to implement or deploy a solution?
   d. Policy

6. Which risk assessment model has a cost involved?
   a. Quantitative

7. A password alone is considered _____.
   b. Single factor

8. What is needed inside a wireless security policy to protect users from the threats of hotspots?
   c. Public access policy

9. Trying all types of password combinations is called a _____.
   b. Brute force attack

10. Which password has three complexity requirements?
    b. 23j@w09

11. Which action will not help the strength of a password?
    a. Limit the frequency of reuse

12. What document is used to instruct the lowest level of management?
    c. Process

13. What document is used to define *how* a policy will be achieved?
    a. Standard

# Chapter 15 Review Answers

1. What is the most inexpensive architecture to deploy, based solely on hardware?
   a. WEP

2. What architecture is most risky in terms of security?
   a. WEP

3. In a VPN, what protocol creates the keys?
   c. ISAKMP

4. When considering a VPN architecture, what two main options do you have to choose from?
   a. Locally placing devices
   b. Remotely placing devices

5. When looking at the 802.1x architecture, what RADIUS server has the cost of licensing along with the cost of the server?
   c. Funk

6. When looking at a wireless gateway, what vendor would be the best to use if a clientless solution is needed?
   c. BlueSocket

7. In a VPN, what protocol exchanges the keys?
   b. IKE

8. When looking at a wireless gateway, what vendor would be the best to use if a high-security, military-grade solution is needed?
   a. Air Fortress

9. What cost savings can be seen when using a Microsoft RADIUS solution if the client has an existing MS network in place?
   a. Hardware
   b. Software
   c. Installation

10. Name one of the largest hidden costs associated with the VPN architecture.
    c. Cost to keep the wireless in a DMZ

11. Which gateway product provides services for worm detection?
    c. Vernier

12. Which IPSec mode is most common for wireless?
    b. ESP

13. Which two architectures require moving the wireless into a DMZ?
    b. VPN
    c. Wireless firewall or gateway

14. Which architecture is the most difficult to change after a single key is compromised?
    a. WEP

## Chapter 16 Review Answers

1. What type of tool is Network Stumbler?
   b. Wireless scanner tool

2. Wireless scanners all perform sniffing functions in the background.
   a. True

3. What type of tool is AiroPeek?
   a. Wireless sniffer tool

4. Which of the following tools are not free of charge?
   c. AiroPeek
   d. AirMagnet

5. What type of tool is AirSnort?
   c. Wireless cracker tool

6. What type of tool is MiniStumbler?
   b. Wireless scanner tool

7. Wireless scanners are used for advanced troubleshooting.
   b. False

8. What type of tool is Kismet?
   e. Wireless hybrid tool

9. What type of tool is AirMagnet?
   e. Wireless hybrid tool

10. Which of the following tools are free of charge?
    a. AirSnort
    b. WEPCrack

11. Which of the following tools do not crack WEP keys?
    d. Kismet

12. What type of tool is WLAN-Jack?
    d. Wireless DoS tool

13. Which feature does AiroPeek and AirMagnet share?
    c. Rogue access point identification

14. Kismet can be used for a Wireless Intrusion Detection System (WIDS).
    a. True

15. On what platforms does Network Stumber work?
    c. Windows 2000
    d. Windows XP

# Index